DOCUMENTING HISTORY

SCIENTISTS
AND THEIR
DISCOVERIES

CHRISTINE HATT

FRANKLIN WATTS
A Division of Scholastic Inc.
NEW YORK TORONTO LONDON AUCKLAND SYDNEY

MEXICO CITY NEW DELHI HONG KONG

DANBURY, CONNECTICUT

First published by Evans Brothers Limited
2A Portman Mansions
Chiltern Street
London, W1M ILE

© Evans Brothers Limited 2001

First American edition 2001 by Franklin Watts
A Division of Scholastic Inc.
90 Sherman Turnpike
Danbury, CT 06816

Catalog details are available from the Library of Congress
Cataloging-in-Publication Data

Printed in Spain by GRAFO

ISBN 0-531-14614-6 (Lib. Bdg.)

Design – Neil Sayer
Editorial – Nicola Barber
Illustrations – Richard Morris
Consultant – Peter R. Ellis
Production – Jenny Mulvanny

Title page picture: Albert Einstein

Border: Microscope and genes

To find out more about the scientists mentioned in this book, visit these Web sites:

http://dir.yahoo.com/science/people/

http://www.pbs.org/wgbh/aso/databank/bioindex.html
(People and Discoveries)

http://www.webelements.com
(for information about the Periodic Table)

ACKNOWLEDGMENTS

For permission to reproduce pictorial material, the author and publishers gratefully acknowledge the following:

Cover: (top left) Science Photo Library (bottom left) Science Photo Library (middle) Science & Society Picture Library (top right) Science & Society Picture Library (bottom right) Science & Society Picture Library **page 8** Science & Society Picture Library **page 9** (top) Science Photo Library (bottom) Science Photo Library **page 10** Science Photo Library **page 11** Science Photo Library **page 12** Science Photo Library **page 13** Topham Picturepoint **page 14** Science Photo Library **page 15** (top) Science & Society Picture Library (bottom) Science & Society Picture Library **page 16** Science Photo Library **page 17** (top) Science Photo Library (bottom) Science Photo Library **page 18** Science Photo Library **page 19** (top) Science & Society Picture Library (bottom) Science Photo Library **page 20** Science Photo Library **page 21** (left) Science & Society Picture Library (right) Science Photo Library **page 22** (top) Science & Society Picture Library (bottom) Science Photo Library **page 23** Mary Evans Picture Library **page 24** Science Photo Library **page 25** Hulton Getty **page 26** Science & Society Picture Library **page 27** Science Photo Library **page 28** Science & Society Picture Library **page 29** (left) Science Photo Library (right) Science Photo Library **page 30** Science Photo Library **page 31** (top) Science Photo Library (middle) Science & Society Picture Library **page 32** Science Photo Library **page 33** (left) Science Photo Library (right) Science & Society Picture Library **page 34** Science & Society Picture Library **page 35** Science & Society Picture Library **page 36** (left) Science Photo Library (right) Science Photo Library **page 38** (left) Science Photo Library (right) Mary Evans Picture Library **page 39** Oxford Scientific Films **page 40** Science Photo Library **page 42** Geology Department, Edinburgh University **page 43** Bruce Coleman Collection **page 45** (left) Photri Microstock (right) Science Photo Library **page 46** Science & Society Picture Library **page 48** Science & Society Picture Library **page 49** (left) Science Photo Library (right) Bruce Coleman Collection **page 50** Science Photo Library **page 51** (top) Mary Evans Picture Library (bottom) Science & Society Picture Library **page 52** (left) Science & Society Picture Library (right) Science Photo Library **page 53** Photri Microstock **page 54** (left) Science & Society Picture Library (right) Science Photo Library **page 55** (top) Science & Society Picture Library (bottom) Science & Society Picture Library **page 56** Science Photo Library **page 57** Science Photo Library **page 58** (top) Science Photo Library (bottom) Science Photo Library **page 59** Topham Picturepoint

For permission to reproduce copyright material for the documents, the author and publishers gratefully acknowledge the following:

page 9 From *Cecilia Payne-Gaposchkin: An Autobiography and Other Recollections*, edited by Katherine Haramundais, published in 1996 by Cambridge University Press, © Cambridge University Press 1984,1996 **page 11** (top) From *The Great Copernicus Chase and Other Adventures in Astronomical History* by Owen Gingerich, published in 1992 by the Sky Publishing Corporation and Cambridge University Press, © Sky Publishing Corporation 1992 **page 11** (bottom) From the *Selected Papers of Edwin Hubble*, from the Edwin Hubble Collection, Huntington Library, San Marino, California **page 12** From *Lemaître, Big Bang and the Quantum Universe* by Michael Heller, published in 1996 by Pachart Publishing House, © 1996 by the Pachart Foundation **page 14** From *The Creation of the Universe* by George Gamow, published by Viking Press, New York, 1952, © Cambridge University Press 1965 **page 15** From *Mr. Tompkins in Paperback* by George Gamow, published in 1965 by Cambridge University Press, © George Gamow 1952 **page 17** From *A Brief History of Time* by Stephen Hawking, published in 1996 by Bantam Press, a division of Transworld Publishers Ltd, © Stephen Hawking 1988, 1996 **page 19** From *A New System of Chemical Philosophy Part 1* by John Dalton, published in 1803 by R. Bickerstaff, London **pages 23, 24, and 25** From *Madame Curie* by Eve Curie, published in 1938 by William Heinemann Ltd **page 27** (middle and bottom) From *Dorothy Hodgkin: A Life* by Georgina Ferry, published in Great Britain by Granta Books 1998, © Georgina Ferry 1998 **page 29** From *Michael Faraday and the Royal Institution* by John Meurig-Thomas, published in 1991 by Adam Hilger, © IOP Publishing Ltd 1991 **page 31** (middle, bottom) From *The Demon in the Aether, The Story of James Clerk Maxwell* by Martin Goldman, published in 1983 by Paul Harris Publishing, © 1983 Martin Goldman **page 35** From *Introducing Einstein* by Joseph Schwartz and Michael McGuiness, edited by Richard Appignanesi, published in 1999 by Icon Books Ltd, © 1979 Joseph Schwartz (text), © 1979 Michael McGuiness (illustrations) **page 39** (middle, bottom) From *Life, Letters and Journals of Sir Charles Lyell, Bart*, edited by Mrs. Lyell, published in 1881 by John Murray, republished 1970 by Gregg International Publishers Ltd **page 41** From *The Origin of Continents and Oceans* by Alfred Wegener, published in 1924 by Methuen & Co Ltd, © 1966 Dover Publications **page 42 and 43** From *The Collected Papers of Lord Rutherford of Nelson*, published in 1963 by George Allen & Unwin Ltd, © 1963 George Allen & Unwin Ltd **page 45** From *Studies in Earth and Space Sciences*, edited by R. Shagam, R.B. Hargraves, W.G. Morgan, F.B. Van Houten, C.A. Burk, H.D. Holland, L.C. Hollister, published by The Geographical Society of America, Inc, © The Geographical Society of America 1972 **page 47 and 49** From *The Origin of Species by Means of Natural Selection* by Charles Darwin, published in 1859 by Penguin Books **page 51** From *Experiments in Plant Hybridization* by Gregor Mendel, edited by J.H. Bennett, published by Oliver & Boyd 1965, © 1965 Oliver & Boyd **page 53** From *Louis Pasteur* by Beverley Birch, published in 1990 by Exley Publications Ltd, © Exley Publications 1990 **page 55** From *Penicillin: Its Practical Application* by Alexander Fleming, published in 1946 by Butterworth & Co **page 57** From *What Mad Pursuit* by Francis Crick, published in 1998 by Basic Books Inc, © 1988 by Francis Crick **page 59** (left) From *The Harrap Science Encyclopedia* (right) from the McGraw-Hill Encyclopedia of World Biography, © 1973 by McGraw-Hill, Inc.

While very effort has been made to secure permission to use copywritten material, Evans Brothers apologizes for any errors or omissions in the above list and would be grateful for notification of any corrections to be included in subsequent editions.

CONTENTS

LOOKING AT DOCUMENTS

This book is about some of the great scientists whose discoveries transformed the world in the nineteenth and twentieth centuries. It looks at their personal backgrounds, the choices they made, and the obstacles they overcame to reach their goals. It examines their ideas and achievements, and how they affected society as a whole. Above all, it highlights their work: the mixture of experiment, collaboration, and brilliant individual insight that has characterized scientific progress through the ages.

The earliest scientists investigated any aspect of the natural world that aroused their interest, from animals to planets. However, by the nineteenth century, scientific knowledge was growing at such a rate that people specialized in particular areas of research. *Scientists and Their Discoveries* reflects this modern approach by dividing its content into five chapters, each dedicated to a different type of scientific expert: astronomers, chemists, physicists, geologists, and biologists.

During the period covered by this book, each of these groups made major advances. Astronomers such as George Gamow developed theories about how the universe began, while chemists such as John Dalton made significant progress toward understanding the nature of matter. Some physicists delved further into this mystery, while others revealed the secrets of forces such as electricity and magnetism. Albert Einstein, the greatest physicist of the era, gained a revolutionary new insight into the natural laws that govern everything in the universe.

Geologists of this period made discoveries that were just as revolutionary. By examining the Earth, Charles Lyell and others overthrew ancient ideas about how landscapes and continents formed. Biologists, too, made a substantial mark. Charles Darwin, for example, developed his far-reaching theory of evolution. In the 1950s, Francis Crick and James Watson discovered the structure of DNA, the basic chemical building block of life.

To bring the stories of these people to life, *Scientists and Their Discoveries* uses a wide range of documents. They include handwritten notebooks, letters of encouragement and congratulation between experts working in the same field, famous books that announced scientific breakthroughs to the general public, papers and magazines designed for specialists, and biographies.

To make the documents easy to read and understand, we have printed them all in the same type. However, you will find photographs of some of the original documents, often including illustrations, alongside several of the excerpts. Difficult or technical words and phrases are explained in the captions around the documents.

On these pages are a few excerpts from the documents used in this book. They have been selected to give you an idea of the great variety of documents included, and to explain how and why some of the documents were written.

Books of popular science, in which experts write about complex ideas in simple terms, make scientific concepts accessible to all readers. In *Mr. Tompkins in Wonderland*, **astronomer and physicist George Gamow brings arguments about the universe to life with drawings and poetry.**

What does this mean? Some words or phrases in the documents are difficult to understand. Captions alongside the documents give explanations of highlighted areas of text. You can find out what this line means on page 15.

My telescope
Has dashed your hope;
Your tenets are refuted.
Let me be terse:
Our universe
Grows daily more diluted!

Scientists often kept handwritten notebooks in which they recorded the details of their experiments. These reveal the basic data on which their theories were built. In this notebook (right), nineteenth-century chemist John Dalton wrote down the results of his tests for color blindness (see page 19).

Scientists also use notebooks to jot down ideas for their theories. Ernest Rutherford drew this diagram of atomic structure (left) in 1910 or 1911 (see page 33). It was based on the results of experiments he had completed in his laboratory at Manchester University.

Biographies often contain information about scientists' personal lives, as well as their careers. For example, Eve Curie's biography of her mother, the great Polish-French chemist Marie Curie, shows her to be a doting parent as well a committed scientist (see page 23).

Books in which great scientists reveal and explain their new ideas are extremely important documents. Among the most influential is *The Origin of Species* by Charles Darwin (see page 49). In 1859, he used this work to announce his theory of evolution to the world.

Irène has cut her seventh tooth, on the lower left...

The various reasons we have just enumerated lead us to believe that the new radioactive substance contains a new element to which we propose to give the name of RADIUM...

ON

THE ORIGIN OF SPECIES

BY MEANS OF NATURAL SELECTION,

OR THE

PRESERVATION OF FAVOURED RACES IN THE STRUGGLE FOR LIFE.

By CHARLES DARWIN, M.A.,
FELLOW OF THE ROYAL, GEOLOGICAL, LINNÆAN, ETC., SOCIETIES;
AUTHOR OF 'JOURNAL OF RESEARCHES DURING H. M. S. BEAGLE'S VOYAGE
ROUND THE WORLD.'

LONDON:
JOHN MURRAY, ALBEMARLE STREET.
1859.

The right of Translation is reserved.

ASTRONOMERS

HENRY RUSSELL
AND THE STUDY OF THE STARS

Until the 1800s, astronomers could do little more than survey the night skies and plot the movements of the stars. The birth of astrophysics—the study of what stars are made of and how they evolve—opened up many possibilities. One of the first great astrophysicists was an American, Henry Norris Russell (1877–1957).

Astrophysics was made possible by spectroscopy. Scientists in the nineteenth century knew that white light was made up of colors, and that it could be split to show this spectrum. In 1814, German physicist Josef von Fraunhofer split light from the sun and found that there were dark lines in the color bands. Later research showed that these lines were caused by different elements in the sun absorbing some of the light. The pattern of the lines gave scientists information about which elements were present in the sun. It became clear that by analyzing the spectra of other stars, scientists could learn what they were made of, too.

By the early twentieth century, astronomers at Harvard University were busy classifying stars according to their spectra. Their work came to the attention of Henry Russell, professor of astronomy at Princeton University. He was especially interested in the life cycles of stars, and thought that the new classifications might help him find out more. In 1913, Russell drew a graph using the Harvard data. He figured out each star's position by plotting its spectrum classification (color) against its luminosity (true brightness). As the star's color revealed its temperature—blue stars are hottest and red coolest—Russell also plotted temperature against brightness. He found that most stars fell in a diagonal from top left to bottom right—the main sequence.

There were exceptions. Some stars appeared on the top right, indicating that they were cool but bright. This meant they must be huge, so that they shone strongly despite their coolness. These stars became known as red giants. Other stars appeared on the bottom left, indicating that

THE HERTZSPRUNG-RUSSELL DIAGRAM

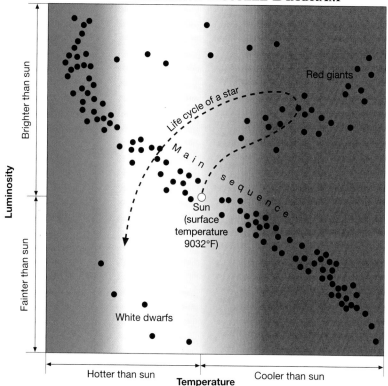

This modern version of the Hertzsprung-Russell Diagram uses color as well as a scale (along the bottom) to indicate the stars' temperatures. The original diagram was in black and white.

they were hot but dim. This was because they were very dense but too small to shine brightly. These stars were called white dwarfs.

Astronomers slowly learned to interpret the graph. They found that a star begins life in the main sequence, where it turns hydrogen into helium to produce energy. When its supply of hydrogen runs out, the star collapses, making its core heat up and its outer layers expand. The layers then cool and redden, forming a red giant. Finally, they drift into space, leaving only a core. The star is now a white dwarf and will soon die.

Russell's graph, now known as the Hertzsprung-Russell Diagram, is a guide to the evolution of stars. It is still widely consulted by astrophysicists around the world.

Henry Russell came from Oyster Bay, New York. He studied at Princeton University, then at Cambridge University in England before returning to teach at Princeton. In 1911, Russell became Princeton's professor of astronomy. Eleven years later, he was appointed research associate at Mount Wilson Observatory in California.

EJNAR HERTZSPRUNG

Henry Russell was not the first person to suggest the existence of main-sequence, giant, and dwarf stars. Danish astronomer Ejnar Hertzsprung had written articles on the subject in a German journal as early as 1905. This is why the graph comparing color with brightness is now known as the Hertzsprung-Russell Diagram. Hertzsprung is also famous for helping devise a new and accurate method of calculating the distances of stars from the Earth.

 Women carried out important early work in astrophysics. Among them was an Englishwoman, Cecilia Payne-Gaposchkin (right), who studied at Cambridge University. She moved to Harvard University in 1923, becoming professor of astronomy in 1956. In this excerpt from her autobiography (1984), she compares her experiences as a female astronomer in England and the United States.

Payne-Gaposchkin is talking about the 1930s here.

Women [at Cambridge] were segregated in the lecture room. Even in the laboratory they were... treated as second-class students... We manage things better in the United States. Even 50 years ago a woman might do astronomical research and even make a name by publication. She might hold a position—without a title and ill-paid, it is true—and she could meet on equal terms with any astronomer in the world. In my early days at Harvard... How we argued, how we walked about the streets and sat talking in restaurants until the manager turned off the lights in despair! We met as equals; nobody ever condescended to me on account of sex or youth.

EDWIN HUBBLE
AND THE EXPANDING UNIVERSE

By the twentieth century, astronomers knew that the Earth was part of the Milky Way galaxy, a vast collection of stars. They were not sure what lay beyond, however. To find out, American astronomer Edwin Powell Hubble uncovered important truths about the nature of the universe.

Hubble originally trained as a lawyer, but in 1914 decided to become an astronomer instead. Eventually, in 1919, he took a job at the Mount Wilson Observatory in California. Another Mount Wilson astronomer, Harlow Shapley, was already investigating the size and shape of the Milky Way. To do so, he had used a new method of calculating distance devised by Henrietta Leavitt. Some experts believed Shapley had mistakenly included objects that were not part of the galaxy in the area he had measured, and therefore his measurements were wrong. In particular, they thought the spiral nebulae —bright, spiral-shaped clouds of gas and dust—he had included were separate galaxies.

Hubble soon began his own observations of the nebulae to see what they really were. He concentrated on the Andromeda Nebula, peering at it through Mount Wilson's new 100-inch (2.54-meter) telescope. It revealed that the "nebula" was in fact made up of stars. In 1924, Hubble showed that Andromeda was far beyond the possible limits of the Milky Way. In other words, it was another, separate galaxy, the first whose existence had been proved.

Hubble soon discovered that more so-called nebulae were actually galaxies. He also noticed that the spectra produced by light from many of them displayed red shift. This means that, compared with a standard light source, the dark lines in their color bands had shifted toward the red end of the spectrum. This effect occurs when light waves from an object lengthen, which happens when the object is moving away from the Earth. The greater the red shift, the faster the object is retreating.

By 1929, Hubble had gathered a mass of information about the distances of galaxies from the Earth and, using red shifts, about their speed. He was able to show that the more distant a galaxy is, the faster it is moving. In 1929, he published

The existence of the Andromeda Galaxy (center) was proved by Edwin Hubble in 1924. Its vast distance from Earth—2.2 million light years—had previously made its identification as a galaxy impossible.

THE SECRETS OF THE CEPHEIDS

The Cepheid variables are a special group of yellow stars. They are so called because the first of these stars was discovered in the constellation Cepheus, and because they pulsate, emitting a variable amount of light. In the early twentieth century, American astronomer Henrietta Leavitt studied these stars. She found that the longer a Cepheid took to complete each pulsation, the greater its luminosity (true brightness). Astronomers can therefore figure out a star's luminosity from its pulsation period. By comparing a star's luminosity with its apparent brightness as seen from Earth, they can also calculate its distance from the Earth. Leavitt's work was further developed by Ejnar Hertzsprung, Harlow Shapley, and others.

Hubble's Law, which expressed this observation in a mathematical formula. His findings proved that the universe is expanding at a rapid rate. They eventually helped other astronomers calculate the age and possible fate of the universe.

RED SHIFT

Spectrum of stationary light source

Spectrum of receding light source

This diagram shows that the light waves from an object lengthen when it is moving away from the Earth, and that the dark bands in the light's spectrum shift toward the red end.

Astronomers of the 1920s were a close-knit community. This excerpt is from a letter that Hubble wrote to Harlow Shapley, by then director of the Harvard College Observatory, when he discovered the first Cepheid variables in the Andromeda Nebula.

In the late eighteenth century, French astronomer Charles Messier (1730–1817) made three catalogs of all the "nebulae" that he could see. He gave each one an "M-number" like this.

EDWIN POWELL HUBBLE (1889–1953)

Edwin Powell Hubble was born in Marshfield, Missouri. He studied law at the University of Chicago, then at Oxford, England. Throughout his student years, he was an amateur astronomer as well as a gifted athlete and boxer. On returning to the United States in 1913, Hubble became a lawyer. However, a year later he abandoned this career to take a job at the Yerkes Observatory, part of the University of Chicago. He moved to the Mount Wilson Observatory in 1919. There, he not only proved that the universe is expanding, but also devised the first major galaxy classification system. Hubble stayed at Mount Wilson for the remainder of his career.

Edwin Powell Hubble in front of the 100-inch (2.54-m) Mount Wilson telescope that he used to study the Andromeda Galaxy

February 19, 1924.

...
Dear Shapley:
    You will be interested to hear that I have found a Cepheid variable in the Andromeda Nebula (M31). I have followed the nebula this season as closely as the weather permitted... in the last five months [and found] two variables... last week... Enclosed is a copy of the normal light curve [of the first], which... shows the Cepheid characteristics in an unmistakable fashion. I have a feeling that more variables will be found... Altogether the next season should be a merry one.

GEORGE GAMOW AND THE BIG BANG

Edwin Hubble showed that the universe is expanding. Other scientists, such as Ukrainian-American George Gamow, went on to consider how and when this ever-changing universe began—and where it might end.

Hubble was the first astronomer to find physical evidence of the universe's expansion, in 1924. At about the same time, other astronomers were trying to find theoretical proof of the same phenomenon. In 1927, Georges Edouard Lemaître, a Roman Catholic priest from Belgium, used Einstein's equations to calculate that the universe was expanding. He then went on to speculate about how that expansion had begun. He theorized that all the matter in the universe had originally been squashed into one incredibly dense "primeval atom." This atom, which he also called the "cosmic egg," had at first slowly disintegrated, then become unstable and violently exploded. As a result, all the matter it contained was still flying apart.

In the 1930s, Lemaître's

Abbé (Father) Georges Lemaître

 After Georges Lemaître died in 1966, a document titled "The Expanding Universe" was found in his belongings. It probably dates from the late 1930s or early 1940s. The document clearly explains Lemaître's ideas about the origins of the universe. It also shows that he had considered whether the universe had been created by God—but he states that astronomy cannot answer such questions.

17. THE BEGINNING OF SPACE.

...Physically everything happens as if [the origin of the universe] was really a beginning. The question if it was really a beginning or rather a creation: something starting from nothing, is a philosophical question which cannot be settled by physical or astronomical considerations.

There is a nice way to obtain a universe formed of fresh matter... I mean the primeval atom hypothesis. The universe would have started as an atom... The atom would break into two parts, equal or unequal, each piece would break again until after some two hundred and fifty such processes, matter would be reduced to its actual... state.

Two pages from Lemaître's annotated manuscript

theory caught the attention of other scientists. Among them was Gamow, who was by then working at George Washington University in Washington, D.C. As Lemaître had provided no real proof of his theory, Gamow set out to do just that. He worked with a student, Ralph Alpher, to carry out his investigations, and in 1948 the two men published their results. They showed that Lemaître had been basically correct—the universe had begun with an explosion. It was caused not by the breakdown of the "primeval atom," but by the gradual build-up of a hot, dense cloud of gas. Gamow called this giant explosion the Big Bang, the name by which it is still known.

The Big Bang theory seemed likely. However, Gamow still had to figure out how the chemical elements that make up the universe could have formed in this process. In the 1950s, he proved how, in the hot, dense ⮑

THE BIG BANG

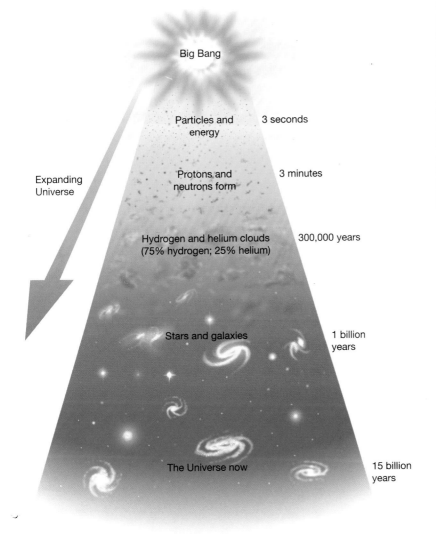

Big Bang

Particles and energy — 3 seconds

Expanding Universe

Protons and neutrons form — 3 minutes

Hydrogen and helium clouds (75% hydrogen; 25% helium) — 300,000 years

Stars and galaxies — 1 billion years

The Universe now — 15 billion years

The Big Bang caused all the matter in the universe, which had been packed into a super-dense mass, to explode outward. Gradually the matter took on new forms until the universe we know today was formed.

GEORGE GAMOW (1904–68)

George Gamow (below) was born in Odessa, Ukraine. He became interested in science as a young child. On his fourteenth birthday, his father gave him a telescope, and he started to observe the skies regularly. After leaving school, Gamow continued his studies at the University of Leningrad in the former Soviet Union before going abroad to carry out research. He met many other brilliant physicists and flourished in the exciting atmosphere of discovery. Gamow studied hard but found plenty of time for play, too. He was well known for his late-night partying and practical jokes. Finally, in 1934, Gamow settled in the United States, becoming professor of theoretical physics at George Washington University in Washington, D.C. In 1956, he moved to the University of Colorado, where he remained until his death.

early universe, subatomic particles could have combined to produce the two lightest elements, hydrogen and helium. He also showed that it was impossible for heavier elements to have emerged in this way. Scientists now believe that other elements formed from fused hydrogen and helium inside stars.

Even stronger evidence for the Big Bang soon appeared. In 1948, Gamow had suggested that the remains of the radiation produced by the explosion must still be in the universe. He had also said that, as the radiation had been cooling ever since, its temperature would now be approximately -518°F (-270°C). In 1964, American astrophysicists Arno Penzias and Robert Wilson detected this background radiation and confirmed the temperature Gamow had predicted. For many scientists, this was the final proof that the universe really did begin with the Big Bang.

In 1989, the Cosmic Background Explorer (COBE) satellite was launched to measure background radiation. Data from the satellite was used to make this microwave map of the universe. The colors show temperature variations. Dark blue areas are at the radiation's average temperature. Light blue areas are cooler, and red and pink areas hotter.

THE STEADY STATE THEORY

In the 1940s, three astronomers came out in strong opposition to Gamow's Big Bang theory: Hermann Bondi, Thomas Gold, and Fred Hoyle. In *A New Model for the Expanding Universe* (1948), Hoyle explained his Steady State theory. It says that the Big Bang never happened, but that the universe is expanding because new matter is continually being created. This new matter replaces the matter drifting away into space, so the universe always looks the same. At first, the Steady State theory had many supporters. Penzias and Wilson provided powerful evidence for the Big Bang, however, and few of today's scientists accept Hoyle's ideas.

Gamow gave the first detailed explanation of his Big Bang theory in *The Creation of the Universe* (1952).

The discovery that our universe is expanding provided a master key to the treasure chest of cosmological riddles. If the universe is now expanding, it must have been once upon a time in a state of high compression. The matter which is now scattered through the vast empty space of the universe in tiny portions which are individual stars must at that time have been squeezed into a uniform mass of very high density... At present the possible maximum density of this... matter is not accurately known. The nearest guess is that... each cubic centimeter of space contained at that time a hundred million tons of matter!

Cosmological means relating to the study of the universe.

Gamow wrote many popular books about science, in which he tried to explain difficult ideas to the general reader. The books featured quirky drawings, some by Gamow himself, as well as songs and poetry. This poem (below right), from *Mr. Tompkins in Wonderland*, contains an imaginary argument between two British astronomers. The first, Martin Ryle, helped prove the Big Bang theory using information gathered by radio telescope. The second, Fred Hoyle, was the main originator of the Steady State theory.

FROM BEGINNING TO END

Using Hubble's Law and his own observations, Gamow calculated that the Big Bang took place about seventeen billion years ago. Modern scientists generally agree. It is impossible to know exactly how the universe will develop in the future, but there are three main options. The universe may be "flat," which means it will use up all its energy and come to a standstill. It may be "closed," which means it will stop expanding and contract until all its matter comes together again in the "Big Crunch." However, the most recent findings suggest that the universe is probably "open," which means it will continue to expand forever.

Part of the poem shown on the right as it appears in Gamow's book

COSMIC OPERA

one million light-years away—show how it actually looked one million years ago. Thus, what Ryle sees, or should I rather say hears, through his radio-telescope, corresponds to the situation which existed in that distant part of the universe many thousand millions of years ago. If the universe were really in a steady state, the picture should be unchanged in time, and very distant galaxies as observed from here now should be seen distributed in space neither more densely nor rarely than the galaxies at shorter distances. Thus Ryle's observations showing that distant galaxies seem to be more closely packed together in space is equivalent to the statement that the galaxies everywhere were packed more closely together in the distant past of thousands of millions of years ago. This contradicts the steady state theory, and supports the original view that the galaxies are dispersing and that their population density is going down. But of course we must be careful and wait for further confirmation of Ryle's results.'

'By the way,' continued the professor, extracting a folded piece of paper from his pocket, 'here is a verse which one of my poetically inclined colleagues wrote recently on this subject.' And he read:

'Your years of toil,'
Said Ryle to Hoyle,
'Are wasted years, believe me.
The steady state
Is out of date.
Unless my eyes deceive me,

My telescope
Has dashed your hope;
Your tenets are refuted.
Let me be terse:
Our universe
Grows daily more diluted!'

Said Hoyle, 'You quote
Lemaître, I note,
And Gamow. Well, forget them!
That errant gang
And their Big Bang—
Why aid them and abet them?

63

THE WOODCARVER

'This is our large cyclotron or "atom-smasher".'

Your tenets are refuted means "your beliefs have been disproved."

See box on page 14.

This lively drawing from another of Gamow's books shows a particle accelerator. A modern example is shown on page 33.

"Your years of toil,"
Said Ryle to Hoyle,
    "Are wasted years, believe me.
The steady state
Is out of date.
    Unless my eyes deceive me,

My telescope
Has dashed your hope;
    Your tenets are refuted.
Let me be terse:
Our universe
    Grows daily more diluted!"

Said Hoyle, "You quote
Lemaître, I note,
    And Gamow. Well, forget them!
That errant gang
And their Big Bang—
    Why aid them and abet them?

You see, my friend,
It has no end
    And there was no beginning,
As Bondi, Gold,
And I will hold
Until our hair is thinning!"

STEPHEN HAWKING AND BLACK HOLES

Toward the end of their lives, many stars turn into red giants, then white dwarfs. However, stars with a mass at least three times greater than that of the sun become black holes. These are among the most mysterious objects in space, but thanks to the work of Stephen Hawking and other scientists, they are slowly revealing their secrets.

In the eighteenth century, astronomers John Michell and Pierre-Simon de Laplace were already wondering if there might be "dark stars" from which light could not escape. As such stars were invisible, these men could not prove their theory. But twentieth-century scientists had other ways of finding out the truth. After Albert Einstein published his general theory of relativity in 1916, scientists used his equations to suggest that, once stars above a certain mass had collapsed, their gravitational fields would prevent light from emerging.

In 1965, British physicist Roger Penrose proved this theory. He showed that these stars collapsed to form a singularity—a point where matter is so dense that it takes up no space. Since even light cannot escape from a singularity, a circular dark area is created around its edge. In 1967, Professor John Wheeler of Princeton University gave a name to this "collapsed star" phenomenon: black hole.

Penrose's research attracted the interest of another physicist, Oxford-educated Englishman Stephen Hawking. Hawking was fascinated by cosmology (the study of the universe), and particularly by the origins of the universe. Research inspired by Penrose's findings led Hawking to write a revolutionary Ph.D. thesis in 1965. It stated that, as a star can collapse inward to form a black hole, a black hole can explode outward. Such a process, Hawking showed, had caused the Big Bang. If it reversed again, it would lead to the Big Crunch.

Since 1965, Hawking has continued to investigate black holes. In particular, he has shown that they are not really "black" at all, as radiation can emerge from them. He has also suggested that dense "mini-black holes" formed just after the Big Bang, but none has yet been detected. Hawking continues to devise complex new theories that are helping unravel the mysteries of the early universe.

STEPHEN HAWKING (1942–)

Stephen William Hawking (below) was born in Oxford, England, and studied at the city's university before going to Cambridge University. In the early 1960s, Hawking started to have difficulty coordinating his movements. Doctors found he had motor neurone disease, which causes muscle wasting and paralysis. Despite this shocking diagnosis, Hawking completed his Ph.D. thesis at Cambridge in 1965. The progress of his disease eventually left him unable to walk or talk, so he began to use a wheelchair and a voice synthesizer to "speak" his words. The career of this highly gifted man nevertheless flourished. In 1979, he became Lucasian Professor of Mathematics at Cambridge, a position he still holds today.

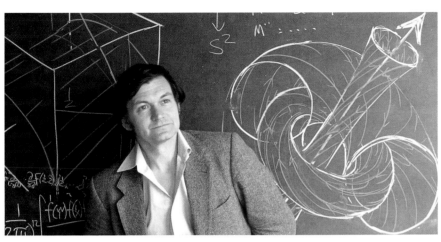

Roger Penrose is Rouse Ball Professor of Mathematics at Oxford University, where he is developing a complex new idea about the universe known as twistor theory. He has also written a book with Stephen Hawking called *The Nature of Space and Time* (1996).

A Black Hole

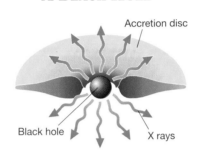

Accretion disc

Black hole

X rays

In 1965, astronomers detected a strong but invisible X-ray source in the constellation Cygnus. Christened Cygnus X-1, it was located near a supergiant star. Hawking believes that the X rays result from the presence of a black hole whose gravity is pulling gas away from the star. This gas forms an accretion disk around the hole (above right). This accretion disk then emits X rays that astronomers can both measure and photograph (right).

A Black Hole Is Born

When a massive star begins to run out of hydrogen, it forms not a red giant, but a red supergiant. Nuclear reactions then take place inside this huge star, forming spheres of different elements around its core. Eventually, most of the core is turned into iron. This causes the star to collapse, then explode as a supernova. In the process, the star's outer layers are hurled into space. If the mass of the remaining core is between 1.4 and 3 times that of the sun, it shrinks to form a dense neutron star. If its mass is any greater, it forms a black hole.

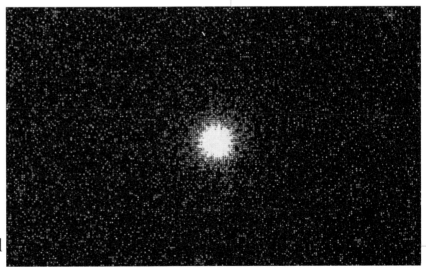

In 1988, Stephen Hawking published *A Brief History of Time*, in which he used simple language to explain cosmology past and present. This excerpt describes what might happen to an astronaut inside a black hole.

If an astronaut falls into a black hole, its mass will increase, but eventually the energy equivalent of that extra mass will be returned to the universe in the form of radiation... Thus, in a sense, the astronaut will be "recycled." It would be a poor sort of immortality, however, because any personal concept of time for the astronaut would almost certainly come to an end as he was torn apart inside the black hole!

CHAPTER 2

CHEMISTS

JOHN DALTON
AND ATOMIC THEORY

Everything—solid, liquid, and gas—is made of matter. Until the seventeenth century, however, scientists did not know what matter itself was made of. John Dalton helped answer this fundamental question.

Ancient Greeks thought matter was made up of four elements—earth, air, fire, and water. Some believed elements were made of atoms (indivisible particles). During the Middle Ages, superstitious alchemists did little to further scientific knowledge. Gradually, scientists began to test their ideas about matter and develop new theories. *The*

Sceptical Chymist (1661), by Irishman Robert Boyle, contained the first modern definition of an element as a substance that cannot be broken down into a simpler one. It also revived the idea that elements were made of "primary particles."

Chemists gradually began to work out which substances were elements according to Boyle's definition and which were compounds—chemically combined groups of elements. For example, in the eighteenth century, English scientist Henry Cavendish showed that water was a compound of the elements

hydrogen and oxygen, and thus could not itself be an element. However, less progress was made in investigating primary particles.

In the nineteenth century, Englishman John Dalton began his studies. Dalton's interest in the makeup of matter was prompted by his study of meteorology. Since weather is affected by the air, he decided to investigate nitrogen and oxygen, the main elements in air. His tests showed these gases were not combined in a compound but in a mixture, where they acted independently. For example, they exerted different amounts of pressure, and one was more easily absorbed by water vapor than the other. Tests on other gases showed that they, too, had different levels of solubility in water—a sign that they also had different weights.

These experiments led Dalton to believe that every element was made up of a different type of particle, each with a different weight. Like the Greeks, Dalton called these particles atoms. In 1803, he announced his ideas to the Manchester Literary and Philosophical Society, and in 1808 published a fuller account in *A New System of Chemical Philosophy*. Dalton's theory, later modified by Amedeo Avogadro, was the foundation of modern science.

JOHN DALTON (1766–1844)

John Dalton (below) was born into a Quaker family in, Cumbria, in the Lake District of England. When he was twelve, he became a teacher at the local school. At fifteen, he took up a post at another school in nearby Kendal, where he also started to teach himself science. In 1793, Dalton became a professor at New College in Manchester, and a year later was elected to the

Manchester Literary and Philosophical Society. The society's members met regularly to give and hear lectures about the latest ideas. Dalton later presented many of his own scientific ideas there, as well as at the Royal Institution in London. He was made a fellow of London's Royal Society Institution in 1822.

COLOR BLINDNESS

John Dalton was color blind, and he carried out many experiments to discover more about this condition. For example, he recorded how silk threads of different colors appeared to him in daylight and in candlelight (right). In 1794, he presented a paper on the subject to the Manchester Literary and Philosophical Society. *Extraordinary Facts Relating to the Vision of Colors* contained the first-ever scientific description of color blindness, now sometimes known as daltonism.

Dalton attached the silks he used to test his color blindness on the left-hand pages of this notebook. He wrote down how the colors appeared to him on the right-hand pages.

The title page of John Dalton's book

In *A New System of Chemical Philosophy* (1808), Dalton explains the essential ideas of his atomic theory. He states that all matter is made of indivisible particles called atoms, and that the atoms of each element are of a different type and atomic weight. He also writes that atoms are rearranged in chemical reactions, and may combine to form compounds. Finally, as this excerpt shows, he declares that atoms cannot be created or destroyed.

This refers to the separation and combination of atoms that take place in chemical reactions.

Chemical analysis and synthesis go no farther than to the separation of particles one from another, and to their reunion. No new creation or destruction of matter is within the reach of chemical agency. We might as well attempt to introduce a new planet into the solar system, or to annihilate one already in existence, as to create or destroy a particle of hydrogen. All the changes we can produce, consist in separating particles that are in a state of cohesion or combination, and joining those that were previously at a distance.

DMITRI MENDELEEV
AND THE PERIODIC TABLE

As the nineteenth century progressed, scientists slowly learned more about atoms. They also discovered more elements until, by the 1860s, they knew of sixty. It was Russian chemist Dmitri Ivanovich Mendeleev who found out how all these elements were related.

In the early nineteenth century, Italian scientist Amedeo Avogadro showed that the smallest naturally occurring particles of an element are not usually individual atoms but molecules. He also showed that the same volume of any gas, at the same temperature and pressure, contains the same number of molecules. This finding, known as Avogadro's Law, provided scientists with a reliable way of calculating the relative weights of molecules and, eventually, of atoms.

Some scientists then suggested that the new atomic weights could be used to classify the elements. In 1864, English chemist John Newlands made one of the first attempts to do this. He noticed that when the known elements were listed in order of their atomic weights, every eighth one had similar properties. He called this the Law of Octaves. Newland's discovery led him to prepare a table in which every eighth element began a new row. The finished layout had seven columns, and the elements in each column shared many important characteristics.

Newlands, however, made one significant error. He failed to consider that some elements might not yet have been discovered—so he left no room for them in his table. It took the insight of Dmitri Mendeleev to take this next step forward. In February 1869, to help order his thoughts for writing a chemistry textbook, he wrote the name of each element and its properties on a card. Then he moved the cards around until he found the layout that best showed the relationships between them. Like Newlands, he put elements with similar properties in the same columns. But if no known element was suitable for a particular slot, he left a gap. Mendeleev also predicted the properties of the elements that he believed would one day fill the gaps.

By the time he had finished, Mendeleev had created the Periodic Table, so called because there are regular periods (intervals) between elements with similar properties. Three of the "missing" elements were soon found, proving the table's value. It has since been much adapted and expanded, and remains a vital scientific tool.

DMITRI IVANOVICH MENDELEEV (1834–1907)
Dmitri Ivanovich Mendeleev (left) was born in Tobolsk, in Siberia, Russia. While he was still a teenager, his father died, so his mother had to bring up her family of fourteen children alone. She wanted Dmitri, her youngest and her favorite, to get a good education, so she sent him to the Central Pedagogic Institute in the city of St. Petersburg. Mendeleev repaid his mother's faith in him when, in 1866, he became professor of chemistry at St. Petersburg University. It was there that he devised the Periodic Table, as well as writing a major textbook, *The Principles of Chemistry* (1868–70). Mendeleev continued to teach at St. Petersburg until 1890, when his support for political change in Russia led him to resign.

The document below is Mendeleev's first, handwritten version of the Periodic Table. The earliest published Periodic Table appeared in *On the Relation of the Properties to the Atomic Weights of Elements* (1869). At the bottom of the page, you can see how the Periodic Table looks today.

CHANGING TABLES

As knowledge grew, the Periodic Table changed. In the twentieth century, scientists discovered that atoms are made up of subatomic particles (see pages 32–3), among them protons. They also found that the atoms of each element contain a different number of protons. Now the elements in the table are arranged according to these atomic numbers, from lowest to highest. The new layout is similar to Mendeleev's because atomic weights, which he used to order his table, rise with atomic number. The number of elements in the table has changed, too. Mendeleev included only about sixty. Now there are at least 118.

American chemist Glenn Seaborg helped discover eight new elements between 1940 and 1958. They included Element 101, which was given the name mendelevium, after Mendeleev.

THE PERIODIC TABLE Each square in this modern Periodic Table contains the atomic number of the element, its chemical symbol, and its full name. Elements that are related, such as particular types of metal or gas, all have squares of the same color.

- Alkali metals
- Alkaline Earth metals
- Transition metals
- Lanthanide metals
- Actinide metals
- Other metals
- Halogens (semimetals)
- Halogens (nonmetals)
- Noble gases

hydrogen 1 H																	helium 2 He
lithium 3 Li	beryllium 4 Be											boron 5 B	carbon 6 C	nitrogen 7 N	oxygen 8 O	fluorine 9 F	neon 10 He
sodium 11 Na	magnesium 12 Mg											aluminum 13 Al	silicon 14 Si	phosphorus 15 P	sulphur 16 S	chlorine 17 Cl	argon 18 Ar
potassium 19 K	calcium 20 Ca	scandium 21 Sc	titanium 22 Ti	vanadium 23 V	chromium 24 Cr	manganese 25 Mn	iron 26 Fe	cobalt 27 Co	nickel 28 Ni	copper 29 Cu	zinc 30 Zn	gallium 31 Ga	germanium 32 Ge	arsenic 33 As	selenium 34 Se	bromine 35 Br	krypton 36 Kr
rubidium 37 Rb	strontium 38 Sr	yttrium 39 Y	zirconium 40 Zr	niobium 41 Nb	molybdenum 42 Mo	technetium 43 Tc	ruthenium 44 Ru	rhodium 45 Rh	palladium 46 Pd	silver 47 Ag	cadium 48 Cd	indium 49 In	tin 50 Sn	antimony 51 Sb	tellurium 52 Te	iodine 53 I	xenon 54 Xe
caesium 55 Cs	barium 56 Ba	57-71	hafnium 72 Hf	tantalum 73 Ta	tungsten 74 W	rhenium 75 Re	osmium 76 Os	iridium 77 Ir	platinum 78 Pt	gold 79 Au	mercury 80 Hg	thallium 81 Ti	lead 82 Pb	bismuth 83 Bi	polonium 84 Po	astatine 85 At	radon 86 Rn
francium 87 Fr	radium 88 Ra	89-103	rutherfordium 104 Rf	dubnium 105 Db	seaborgium 106 Sg	bohrium 107 Bh	hassium 108 Hs	meitnerium 109 Mt	unnnnilium 110 Unn	unununium 111 Uuu	unununbium 112 Uub		ununquadium 114 Uuq		ununhexium 116 Uuh		ununoctium 118 Uuo

lanthanum 57 La	cerium 58 Ce	praseodymium 59 Pr	neodymium 60 Nd	promethium 61 Pm	samarium 62 Sm	europium 63 Eu	gadolinium 64 Gd	terbium 65 Tb	dysprosium 66 Dy	holmium 67 Ho	erbium 68 Er	thulium 69 Tm	ytterbium 70 Yb	lutetium 71 Lu
actinium 89 Ac	thorium 90 Th	protactinium 91 Pa	uranium 92 U	neptunium 93 Np	plutonium 94 Pu	americium 95 Am	curium 96 Cm	berkelium 97 Bk	californium 98 Cf	einsteinium 99 Es	fermium 100 Fm	mendelevium 101 Md	nobelium 102 No	lawrencium 103 Lr

MARIE CURIE
AND RADIOACTIVITY

Several of the elements discovered shortly after Mendeleev's creation of the Periodic Table shared a special property. This was radioactivity, the emission of particles or rays caused by the spontaneous disintegration of atomic nuclei (see page 32). Numerous scientists investigated this phenomenon. Among the most dedicated was a young Polish woman, Marie Sklodowska Curie.

In 1895, German physicist Wilhelm Röntgen discovered X rays (see page 30). His work was noticed by French scientist Antoine Henri Becquerel, professor of physics at the Ecole Polytechnique in Paris. Becquerel was already studying fluorescence, the radiation given off by some substances when they are exposed to the sun's rays. He now began to wonder if the fluorescence from one of these substances, potassium uranyl sulphate, might also contain X rays.

In 1896, Becquerel tested his theory. Knowing that X rays discolored photographic plates, he placed some potassium uranyl sulphate crystals in the sunshine to make them fluoresce, then put a photographic plate wrapped in black paper next to them. He knew that the sun's rays could not get through the black paper, but that any X rays given off by the crystals could. When the photographic plate darkened, he guessed this was a result of the X rays, which were therefore present in the fluorescence.

Then something unexpected happened. The weather grew cloudy, and Becquerel had to delay his experiments. He put the wrapped photographic plate and crystals away in a drawer. When he went to retrieve them, he found that the plate was blurred, even though the crystals had not been fluorescing. Becquerel eventually concluded that another form of

MARIE CURIE (1867–1934)

Marie Curie was born in Warsaw, Poland. In 1891, she left Poland to study science at the Sorbonne, part of Paris University. Four years later, she married another scientist, Frenchman Pierre Curie. In 1904, a year after the couple had won the Nobel Prize, Pierre was appointed professor of physics at the Sorbonne. Marie meanwhile looked after her two daughters and worked part-time at a girls' school. Tragically, in 1906, Pierre was knocked over and killed by a cart. Marie was then appointed to his former position, becoming the first woman professor at the Sorbonne. In 1914, she became head of the Paris Institute of Radium. Unaware of the dangers of radioactivity, Marie never protected herself from the radioactive elements with which she worked. The prolonged exposure eventually caused her to develop leukemia, from which she died.

This illustration of Pierre and Marie Curie appeared on the cover of a French magazine in 1904.

radiation was being produced. He suggested the source was uranium, an element in the crystals.

Now Marie Curie, who was also living in Paris, began to take an interest. By the time Becquerel's findings became public, she was about to begin research for her thesis. The intriguing new form of radiation that Becquerel had discovered was an

Marie Curie's older daughter, Irène, became a scientist like her mother. Marie's younger daughter, Eve, wrote her biography, *Marie Curie* (1938). The two excerpts below show how Marie juggled her work as a scientist with her life as a mother. The first, from her personal notebook, is dated August 1898. The second, from the proceedings of the French Academy of Science, appeared in December of the same year.

Irène has cut her seventh tooth, on the lower left. She can stand for half a minute alone. For the past three days we have bathed her in the river. She cries, but to-day (fourth day) she stopped crying and played with her hands in the water. She plays with the cat and chases him with war cries. She is not afraid of strangers any more. She sings a great deal...

The various reasons we have just enumerated lead us to believe that the new radioactive substance contains a new element to which we propose to give the name of RADIUM. The new radioactive substance certainly contains a very strong proportion of barium; in spite of that its radioactivity is considerable. The radioactivity of radium, therefore, must be enormous.

ideal subject. Curie began by measuring the rays given off by uranium. She then tested other elements and found that one, thorium, emitted the same type of radiation. Curie gave this radiation a name—radioactivity. Next, she tested uranium and thorium compounds. The amount of radioactivity emitted was not affected by the other elements in them, or by varying conditions. It depended only on the amount of uranium or thorium present. She concluded that radioactivity was a property of their atoms. Later investigations proved her right.

Curie's next step was to investigate pitchblende, a uranium-rich mineral. When she measured ➔

WAR WORK

From 1914 to 1918, World War I raged across large areas of western France. In this time of need, Marie Curie put her scientific knowledge to practical use. Realizing that X rays would be a great help to surgeons treating wounded soldiers, she collected X-ray machines from university laboratories and sent them to hospitals in and around Paris. It quickly became clear that doctors would need X-ray equipment in the war zones, too. So, Curie installed the necessary apparatus into about twenty cars, which became known as "little Curies." One of them often carried Curie herself out to help in the battlefield.

A cutaway image of a French X-ray van, showing the special equipment it carried

Barium is a nonradioactive metallic element.

its radioactivity, she found that it was higher than that of pure uranium, so pitchblende had to contain at least one other intensely radioactive element. Her husband, Pierre, abandoned his own work to help his wife search for this substance. In 1898, the couple found two new radioactive elements. The first was named polonium, after Marie's homeland. The second, even more radioactive, was christened radium.

The Curies now set about processing several tons of pitchblende, with the goal of extracting pure radium. They set up a laboratory in a schoolyard shack and worked doggedly for four years. By 1902, they had just one-tenth of a gram, but this was enough for another chemist to figure out its atomic weight. In 1903, Marie finished her thesis. That same year, she, her husband, and Becquerel won the Nobel Prize in physics, for their combined studies of radioactivity. In 1911, Marie alone won

the Nobel Prize in chemistry, for her discovery of radium and polonium.

Marie Curie's discovery of radioactivity has made a lasting impact, both positive and negative. Some of the major benefits are medical applications. For example, radioactive substances can be injected into patients and tracked through their ➲

This haunting image shows Marie Curie at work. She achieved her results with basic apparatus rather than the vast array of complex equipment available to modern scientists.

EXTRA ELEMENTS

In the early twentieth century, scientists following in Marie Curie's footsteps discovered the existence of several more radioactive elements. In 1900, Frenchman André Debierne, once the Curies' assistant at the Sorbonne, found actinium in the mineral pitchblende. German physicist Friedrich Dorn isolated radon, a radioactive gas, in the same year. In 1917, English and German research teams independently discovered another element, protactinium, which was also found in pitchblende.

In the past, as now, the price of fame was often a loss of privacy. In this letter, also taken from *Marie Curie*, Pierre Curie tells a friend how his life has changed since the discovery of radium.

...You have seen this sudden fad for radium. This has brought us all the advantages of a moment of popularity; we have been pursued by the journalists and photographers of every country on earth; they have even gone so far as to reproduce my daughter's conversation with her nurse and to describe the black-and-white cat we have at home. Then we have received letters and visits from all the eccentrics, from all the unappreciated inventors... We have had a large number of requests for money. Last of all, collectors of autographs, snobs, society people and sometimes even scientists come to see us... With all this, there is not a moment of tranquillity in the laboratory...

bodies so that doctors can find out what is wrong, and radioactive rays can destroy cancerous tissue. However, experts have also learned that exposure to radioactivity can kill—Curie herself was a victim. As a result, many people now consider other applications, such as the use of radioactive fuels in nuclear power stations, too dangerous. Scientists everywhere are learning to handle radioactivity with care.

The Paris Institute of Radium in 1955. This picture was taken more than twenty years after the death of its remarkable founder.

The Institute of Radium in Paris opened in 1914. It had two laboratories, one for academic study, the other for research into Curietherapy—medical treatments using radioactivity. Unfortunately, it had very little radium. In 1920, Marie Curie explained the problem to a visiting American, Mrs. William Meloney. Back at home, Mrs. Meloney raised funds to buy a precious gram of radium for the institute. She was so successful that, in 1921, Curie was able to collect the radium from U.S. president Warren G. Harding. This excerpt, from Eve Curie's biography of her mother, is Mrs. Meloney's own account of her first meeting with Marie Curie.

I tried to explain that American women were interested in her great work, and found myself apologizing for intruding upon her precious time. To put me at my ease Mme Curie began to talk about America.

"America," she said, "has about fifty grams of radium. Four of them are in Baltimore, six in Denver, seven in New York." She went on, naming the location of every grain.

"And in France?" I asked.

"My laboratory has hardly more than a gram."...

"If you had the whole world to choose from," I asked impulsively, "what would you take?"

It was a silly question perhaps, but as it happened, a fateful one.

DOROTHY HODGKIN
AND X-RAY CRYSTALLOGRAPHY

In the early twentieth century, scientists were struggling to find out more about atoms and molecules. Their biggest problem was not being able to see the tiny particles. Then, in 1912, a method of examining them using X rays was devised. This technique was put to brilliant use by British chemist Dorothy Crowfoot Hodgkin.

X rays were discovered in 1895. In 1912, as part of his work to reveal their precise nature, German physicist Max von Laue fired X-ray beams into crystals. (Crystals are solids with their atoms arranged in lattices—see diagram.) Von Laue knew that the X rays would hit the atoms and be diffracted (scattered). But he was thrilled to find that, because of the special arrangement of crystal atoms, the scattered rays made regular patterns on photographic plates. Two British scientists, William Henry and William Lawrence Bragg, found that the patterns produced by the X rays were a guide to the three-dimensional structures of the crystals. Bragg figured out a means of interpreting them, and X-ray crystallography was born.

Crystallography developed steadily over the following years, and young Dorothy Hodgkin made it her specialty. Then, in the mid-1930s, Hodgkin used the technique to investigate insulin. This protein, discovered in 1921, is used in the treatment of diabetes, a disorder caused by the body's inability to process carbohydrates properly. Doctors wanted to understand more about the structure of insulin, and Hodgkin produced crystals suitable for analysis in 1935. It was many years before she understood their internal form, however.

During World War II, Hodgkin investigated penicillin, which by the 1940s was in use as an

DOROTHY CROWFOOT HODGKIN (1910–94)

Dorothy Hodgkin (below) was born in Cairo, Egypt. When World War I broke out in 1914, she returned to England with her mother. During the early 1920s, she read two children's lectures about X-ray crystallography by William Henry Bragg, and her lifelong interest began. In 1928, Crowfoot went to Somerville College at Oxford to study chemistry. After graduating, she researched crystallography at Cambridge University, before returning to Oxford as a lecturer. In 1937, she married Africa expert Thomas Hodgkin. Dorothy Hodgkin left Oxford in 1970 to become chancellor of Bristol University.

A SODIUM CHLORIDE CRYSTAL

Chlorine atoms — Sodium atoms

This diagram of a sodium chloride (common salt) crystal shows clearly how the atoms of the two elements are arranged in a lattice.

antibiotic. She revealed its structure in 1945. Three years later, she began to examine Vitamin B-12. This was also of medical interest, as people who cannot absorb this vitamin develop anemia, a serious red blood cell deficiency. Computers helped create a picture of the highly complex molecules that make up B-12. It was ready in 1956, and Hodgkin won the Nobel Prize in chemistry for her work in 1964. Hodgkin and her research team made one more breakthrough. In 1969, after thirty-four years, they finally discovered the structure of insulin.

INTERNATIONAL UNDERSTANDING

Four of Dorothy Hodgkin's uncles were killed in the World War I, and she also witnessed the sufferings of many people in World War II. Hodgkin became dedicated to world peace, and soon joined the Campaign for Nuclear Disarmament. She promoted international understanding by visiting scientists in many countries. In China, she was delighted to be shown an insulin structure map made in 1971, just two years after she had completed her own.

William Henry Bragg and his son shared the Nobel Prize in physics in 1915.

Lecture series for children were regular events at the Royal Institution in London in 1826. In 1923 and 1925, the lectures were given by William Henry Bragg.

The lectures set young Hodgkin on her future career path. This description of X-ray crystallography comes from Hodgkin's 1925 lecture, "Concerning the Nature of Things."

The discovery of the X ray has provided means by which we can look far down into the structure of solid bodies, and observe in detail the design of their composition... How far our new powers will carry us, we do not yet know; but it is certain that they will take us far and give us a new insight... Broadly speaking, the discovery of X rays has increased the keenness of our vision a thousand times, and we can now "see" the individual atoms and molecules.

The work of an X-ray crystallographer is difficult and time-consuming. The achievement of a final result can be exhilarating, as this quotation shows. It is part of an article by Hodgkin that appeared in the *British Medical Journal* in 1971.

X rays are diffracted by subatomic particles called electrons. By looking at the patterns the diffractions produce, crystallographers can see how densely the electrons are packed together and thus identify the positions of atoms.

I used to say that the evening I developed the first X-ray photograph I took of insulin in 1935 was the most exciting moment of my life. But the Saturday afternoon in late July 1969, when we realized the insulin electron density map was interpretable, runs that moment very close.

CHAPTER 3

PHYSICISTS
MICHAEL FARADAY
AND ELECTROMAGNETIC INDUCTION

By 1800, scientists knew several basic facts about electricity. They were aware that it could be static or flow in a current, and that it had a positive or negative charge. But understanding of what electricity was, or how its power might be harnessed, was limited. Then, in the 1820s, Englishman Michael Faraday began to find out more.

Other scientists had already begun to investigate electricity in earnest. In 1820, Danish physicist Hans Christian Oersted had shown that when a magnetic compass is placed near a wire with electricity flowing through it, the compass needle moves. This indicated that electric currents produce a magnetic field. French scientist André Marie Ampère went on to discover much more about the link between electricity and magnetism.

Faraday made his first major discovery in 1821, when he suspended a current-carrying wire above a fixed magnet. The magnetic field caused the wire to rotate. Faraday had discovered the motor effect. This led to the development of the electric motor, which today powers everything from hair dryers to cars. Faraday, however, was more interested in his theory that, just as an electric current can produce a magnetic field, a magnetic field can produce an electric current. In 1831, he set up an experiment to test his idea. First, he wound two wire coils around an iron ring, then joined one coil to a battery and the

MICHAEL FARADAY (1791–1867)

Michael Faraday was born in Newington, near London, England. He left school at thirteen and was employed as an apprentice bookbinder. Eager for knowledge, Faraday read parts of the books he bound, including an article on electricity in the *Encyclopedia Britannica*. He was fascinated by what he learned. In 1812, he went to four lectures at the Royal Institution, a scientific organization in London. The lecturer was famous chemist Humphry Davy. Davy's lectures inspired Faraday to pursue a scientific career. So, in 1813, Faraday got a job as one of Davy's assistants at the Royal Institution. In 1825, he became the institution's laboratory director and, in 1833, its professor of chemistry. Faraday's work in the field of chemistry was significant. He discovered benzene in 1825 and published his laws of electrolysis during the 1830s. However, his research into electricity was even more important. Faraday worked in his laboratory at the Royal Institution until his death.

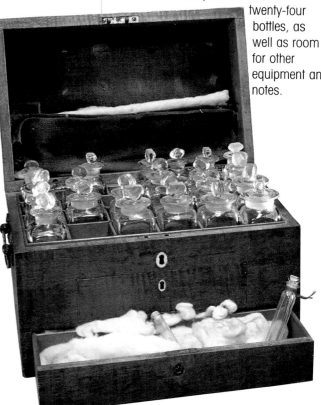

The wooden chemical chest used by Michael Faraday. It had slots for twenty-four bottles, as well as room for other equipment and notes.

other to a galvanometer (an instrument for measuring electric current).

The result was clear. When the battery was connected, electricity flowed in both coils. There could be only one explanation. The battery had made a current flow in the coil to which it was joined. The magnetic field created by this current had then produced a current in the other coil. Faraday's theory had been correct. He later showed that a current could be created by simply moving a magnet in and out of a wire coil. This effect, electromagnetic induction, is how power-station generators now produce electricity. Without it, our modern world could not exist.

FORCE FIELDS

Faraday proposed an important theory about how magnetism and electricity work and affect each other. In 1831, he suggested that magnets were surrounded by invisible lines of force that had an effect on objects at a distance. In 1837, he proposed that current-carrying wires were also surrounded by such lines, or fields, of force. James Clerk Maxwell (see pages 30–31) used advanced mathematics to develop Faraday's ideas.

Michael Faraday gives one of his Christmas lectures to an audience of curious children—and a few adults—at the Royal Institution.

In the 1820s, Faraday organized talks especially for non-scientists at the Royal Institution. He also founded the Royal Institution Christmas lectures for children. The best-known Christmas lecture series that Faraday gave was "The Chemical History of a Candle" (1860). This excerpt is taken from the introduction to the first of these six talks.

Natural philosophy is an old term for physical sciences, or sciences that do not deal with living things.

Faraday is saying that careful examination of even a simple burning candle can open up the whole world of science to a child.

By **seniors**, Faraday means adults. He is telling the children ("juveniles") in his audience that he is speaking to them like a child himself, because he shares their childlike enthusiasm for the subject.

There is no better, there is no more open door by which you can enter into the study of natural philosophy, than by considering the physical phenomena of a candle... before proceeding, let me say this also—that though our subject be so great, and our intention that of treating it honestly, seriously, and philosophically, yet I mean to pass away from all those who are seniors amongst us. I claim the privilege of speaking to juveniles as a juvenile myself...

JAMES CLERK MAXWELL AND ELECTROMAGNETIC THEORY

James Clerk Maxwell (1831–79) was a brilliant Scottish physicist. He made great advances in many fields, particularly the study of gases, color vision, and optics. His most revolutionary work transformed scientific understanding of electricity, magnetism, and light.

As a young science student and lecturer, Maxwell studied the work of Michael Faraday, especially his lines of force theory, for some years. In 1856, he finished a paper that explained the results of Faraday's electromagnetic experiments in mathematical terms. By 1860, when he became professor of natural philosophy and astronomy at King's College, London, Maxwell was ready to continue his research in this field.

Maxwell assembled his theories over the next four years. Then, in 1864, he published his greatest work, *A Dynamical Theory of the Electromagnetic Field*. It contained four equations, known as Maxwell's equations, that explained almost everything about electricity, magnetism, and the relationship between them. In particular, the equations showed that electric and magnetic fields move along together as electromagnetic waves, a type of radiation. The fields occur at right angles to each other, but cannot be separated (see diagram).

The book contained other groundbreaking ideas, too. Maxwell had discovered that electromagnetic waves travel at the speed of light (about 186,000 miles per second—300,000 km per second). He concluded that light itself was a type of electromagnetic radiation. Maxwell's calculations also showed the likelihood of other forms of electromagnetic radiation with different wavelengths. He was proved correct after his death, when radio waves, X rays, gamma rays, ultraviolet rays, and microwaves were all discovered. Together with the previously discovered infrared waves, these make up the electromagnetic spectrum.

Maxwell's theoretical genius has had numerous practical effects that he could never have predicted. Today, gamma rays are used for some types of cancer therapy, while X rays allow doctors to see inside the body. Ultraviolet rays from tanning beds produce artificial tans, although skin experts now consider this to be dangerous. Infrared photography is used for medical diagnosis, the location of underground minerals, and much more. Microwave ovens cook food quickly and efficiently, while radio waves carry sound and television pictures into our homes.

In 1895, German physicist Wilhelm Röntgen was investigating cathode rays by projecting them onto a screen. He noticed that another, unknown form of rays was causing a more distant screen to fluoresce. He had discovered X rays, which in 1912 were shown to be a form of electromagnetic radiation. Doctors now use X rays to produce internal photographs of the body.

AN ELECTROMAGNETIC WAVE

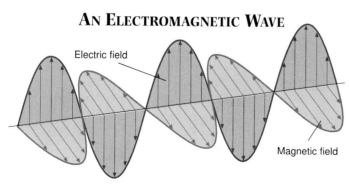

Electric field

Magnetic field

In electromagnetic waves, electricity and magnetism travel along at right angles to each other.

RADIO WAVES

In the 1880s, German physicist Heinrich Hertz put Maxwell's electromagnetic theories to the test. Maxwell had suggested that electromagnetic waves could be produced by setting up a vibrating electric charge. Hertz used a special electric circuit to do just that. The electrical effect on another, unconnected circuit proved that waves had been generated, and that they had traveled through the air. The waves moved at the speed of light, but had a longer wavelength. As Maxwell had predicted, a new form of electromagnetic wave had been discovered. These waves are now known as radio waves.

James Clerk Maxwell

Clarendon Press Series

A TREATISE

ON

ELECTRICITY AND MAGNETISM

BY

JAMES CLERK MAXWELL, M.A.

LL.D. EDIN., F.R.SS. LONDON AND EDINBURGH
HONORARY FELLOW OF TRINITY COLLEGE,
AND PROFESSOR OF EXPERIMENTAL PHYSICS
IN THE UNIVERSITY OF CAMBRIDGE

VOL. I

Oxford

AT THE CLARENDON PRESS
1873

[All rights reserved]

The title page of *A Treatise on Electricity and Magnetism*, Maxwell's second major work, published in 1873

 Maxwell sent his 1856 paper on electromagnetism to Michael Faraday. The following excerpt is taken from Faraday's reply.

Albemarle Street in London is still the head-quarters of the Royal Institution, where Faraday worked.

This means "it is work that I am grateful for."

Albemarle Street, W.,
25th March 1857

MY DEAR SIR—I received your paper, and thank you very much for it. I do not say I venture to thank you for what you have said about "Lines of Force," because I know you have done it for the interests of philosophical truth; but you must suppose it is work grateful to me, and gives me much encouragement to think on. I was at first almost frightened when I saw such mathematical force made to bear upon the subject, and then wondered to see that the subject stood it so well...

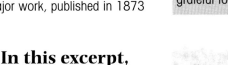 **In this excerpt, Maxwell seems to be describing bells in a belfry. In fact, he is talking about electricity and magnetism. He explains that these forces may appear to act independently, but they are really inseparable.**

In an ordinary belfry, each bell has a rope which comes down through a hole in the floor to the bellringer's room. But suppose that each rope, instead of acting on one bell, contributes to the motion of many pieces of machinery, and that the motion of each piece is determined not by the motion of one rope alone, but by that of several, and suppose, further, that all this machinery is silent and utterly unknown to the men at the ropes, who can only see as far as the holes in the floor above them.

ERNEST RUTHERFORD
AND SUBATOMIC PARTICLES

John Dalton believed atoms were indivisible particles. In the late nineteenth century, however, this theory was gradually demolished by a number of physicists, including the great Ernest Rutherford (1871–1937).

By the mid-nineteenth century, many scientists were using the new technique of spectroscopy to analyze gases. To do so, they had to make the gases glow so that the light from them could be examined. They achieved this effect by putting the gases one by one in a special glass tube, then passing electricity through them. The first unexpected result of such experiments was the discovery of X rays in 1895. Further intriguing developments soon followed.

Joseph John Thomson, professor of experimental physics at Cambridge University, investigated another type of ray that appeared in the gas tests. These rays emerged from the cathode in the glass tube, so they were called cathode rays. By 1897, Thomson had proved cathode rays were made of negatively charged particles. Later, he calculated that each particle's mass was far smaller than the mass of any atom. The tiny new particles were named electrons. Thomson suggested that electrons were stuck in the positively charged bodies of atoms like plums in a pudding.

New Zealander Ernest Rutherford, a former student of Thomson's, shed more light on atomic structure during his investigations of radioactivity. He showed that there were at least two types, which he called alpha and beta particles, and that both were emitted by atoms as they spontaneously disintegrated.

In 1910, while a professor at Manchester University, Rutherford instructed two of his assistants to fire alpha particles at gold foil. The results were astonishing. Most particles went straight through the foil, suggesting that the atoms in it were made up of empty space and of electrons with little mass that let particles through easily. However, about one particle in eight thousand was strongly deflected. This suggested that the atoms in the foil also had tiny, dense cores, which caused particles that hit them to bounce back. In 1911, Rutherford proposed an atomic structure based on a positively charged center, now called the nucleus, with negatively charged electrons in orbit around it. In 1913, Danish physicist Niels Bohr devised an improved version of this model.

Back at Cambridge in 1919, Rutherford carried out another major experiment: bombarding nitrogen with alpha particles. This caused the gas's atomic nuclei to disintegrate and emit positively charged particles that Rutherford called protons. Protons were shown to be present in the nuclei of all atoms, lending support to Rutherford's model of atomic structure. In 1932, British physicist James Chadwick showed there was another type of particle in the nuclei of all elements except hydrogen. This was the neutron, which has no electric charge.

These ideas of atomic structure allowed scientists to explain many of the properties of elements. But there was much more to discover.

Ernest Rutherford (above right) in the Cavendish Laboratory at Cambridge University

PILES OF PARTICLES

Since the 1950s, machines called particle accelerators have been used to break up atomic nuclei. In this way, scientists have discovered many new types of subatomic particles. They now know of over three hundred, which can be divided into two groups: elementary particles and hadrons. Elementary particles, which include quarks and leptons, cannot be split any further. The electron is a type of lepton. Hadrons can be split into quarks. Protons and neutrons are types of hadrons. One of the most important scientists to investigate elementary particles was Murray Gell-Mann. He was the first to suggest the existence of quarks, in 1964, and he gave them their unusual name.

INSIDE ATOMS

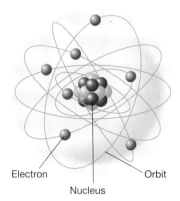

Rutherford model of an atom

Electron   Orbit
Nucleus

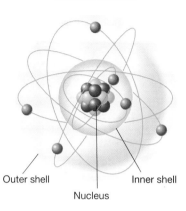

Revised Rutherford-Bohr model of an atom

Outer shell   Inner shell
Nucleus

The diagram on the left shows an atom with electrons moving in fixed orbits around the nucleus, as Rutherford proposed. Scientists now believe that electrons do not move along such fixed paths, but can be found anywhere within broad electron shells, as shown on the right.

 Rutherford probably wrote this page in his notebook during the winter of 1910–11. Headed "Theory of Structure of Atom," it shows the first, rough diagram he made of his new atomic model.

The Tevatron particle accelerator at the Fermi Lab, near Chicago. Particle accelerators are huge machines that hurl elementary particles at very high speeds, often causing them to split and produce new particle types. This particle accelerator is the largest in the world.

Rutherford's diagram is accompanied by equations and notes that provide support for his new theory of the atom.

ALBERT EINSTEIN
AND RELATIVITY

German-born physicist Albert Einstein was the greatest scientist of the twentieth century. Unfortunately, his famous theories of relativity, and many of his other ideas, can be extremely difficult for non-specialists to grasp.

Einstein did not excel in his early years at school. Nevertheless, he continued with his studies and, in 1905, received a doctorate in science from the University of Zurich. In the same year, he published four brilliant papers in a journal called *Annalen der Physik* (*Annals of Physics*), and his

genius emerged. One of these papers, "On the Electrodynamics of Moving Bodies," introduced the Special Theory of Relativity. This was Einstein's solution to a problem that had long been on his mind. This problem was an apparent conflict between two major theories.

The first theory, formulated by Isaac Newton in the seventeenth century, stated that all speeds are relative. For example, if a car is traveling at 50 mph (80 km/h), it is moving at exactly that speed relative to a stationary observer. But relative to

ALBERT EINSTEIN (1879–1955)

Albert Einstein (below) was born into a Jewish family in Ulm, Germany. In 1880, his family moved to the city of Munich. Here Einstein began his schooling, but his progress was slow compared with that of other students. As he moved into his teenage years, Einstein found the rigid school structure increasingly distracting. In 1894, his parents moved once more, this time to Milan, Italy. Einstein stayed behind, but he was expelled from school. He eventually finished his secondary education at Aarau in Switzerland and went on to the Federal Institute of Technology in Zurich. After graduating, Einstein was not given a position at the institute as he had hoped, so he became a private tutor. In 1902, he got a job at the Swiss Patent Office in Bern, while working toward a Ph.D. at the University of Zurich and thinking about physics problems.

The year 1905, when he finished his doctorate and published four major scientific papers, was a turning point in Einstein's life. His work quickly brought him recognition—and job offers. Eventually, in 1914, he settled at the Kaiser Wilhelm Institute for Physics in Berlin. Einstein remained in Germany until 1933, when the Nazis, who persecuted Jews, came to power. They forced him to leave his job, so he moved to the United States, where he became professor of mathematics at Princeton University. During the later years of his life, Einstein campaigned for pacifism and Zionism, the movement for the creation of a Jewish state in Palestine.

another car traveling toward the first one at 37 mph (60 km/h), it is approaching at 87 mph (140 km/h). The second theory, formulated by James Clerk Maxwell, was that the speed of light never changes. The speed at which the light source or the person observing the light is moving makes no difference.

In his Special Theory, Einstein stated that Maxwell was correct. Newton's method of calculating relative speeds would, he predicted, break down as objects approached the speed of light. Nothing, Einstein believed, could travel faster than the speed of light. The combined speeds of two objects moving toward each other could therefore never exceed it, no matter how fast they were going. The closer they got to the speed of light, the more meaningless the calculation would become.

Einstein's theory also stated that as an object approaches the speed of light, its length decreases while its mass increases. A clock attached to the object will also go more slowly the faster the object travels, a phenomenon called time dilation. These effects happen because measurements of space and time are relative, not absolute, as once believed. They change according to the context, like speeds. Scientists have proved these ideas by ➲

English physicist Isaac Newton (1642–1727) deduced the existence of gravity and devised three laws of motion that explain how forces act on objects to produce or prevent movement. Until Albert Einstein introduced the concept of relativity, Newton's ideas were considered to provide a complete picture of the forces at work in the universe.

SEEING THE LIGHT

A second paper published by Einstein in 1905 contained an idea just as revolutionary as the theory of relativity. At this time, scientists were puzzled by the discovery that when light is directed at some metals, they emit electrons. Even top physicists did not understand how light waves could produce this photoelectric effect. In his paper, Einstein showed that light existed and acted as particles, as well as waves, and that the particles' special properties accounted for the effect. He called these particles quanta; they are now known as photons. This new understanding of light helped transform physics.

Even Einstein found his own theories difficult to comprehend at first! This is what he said about the period of his life when he was formulating the Special Theory of Relativity.

I must confess that at the very beginning when the Special Theory of Relativity began to germinate in me, I was visited by all sorts of nervous conflicts. When young I used to go away for weeks in a state of confusion, as one who at that time had yet to overcome the state of stupefaction in his first encounter with such questions.

Stupefaction means astonishment and confusion.

sending tiny particles around particle accelerators at great speeds.

The Special Theory applied only to objects that were stationary or moving at fixed speeds. After his theory's publication, Einstein worked on an extension that would deal with accelerating objects, too. This was published as *The Foundation of the General Theory of Relativity* in 1916. The General Theory is a complex piece of mathematics, but one of its central ideas can be explained more simply. Isaac Newton had taught that gravity is a force operating between all bodies with mass, pulling them toward one another. The more massive the object, the greater the force it exerts. Einstein stated that gravity is not a property of mass, but of space. Massive bodies distort space so that it is curved, as if you dropped a ball into a flat rubber sheet. Less massive bodies, as well as light rays, "fall" into the apparent hollow.

Despite his brilliance, Einstein did make some errors. Using his own General Theory equations, he calculated that the universe was not expanding but static. He later described this conclusion, corrected during his own lifetime by Edwin Hubble and others, as "my biggest mistake." Einstein also stumbled over quantum theory, which was developed by German physicist Max Planck in 1900. It states that matter emits and absorbs energy in tiny units called quanta. Einstein himself successfully applied the theory to light. However, later developments led him to believe that it was flawed. The theory has since been generally accepted.

These lapses did little to damage Einstein's reputation, which had grown quickly following the publication of the General Theory. His worldwide fame was assured when the 1919 solar eclipse proved his theory. Einstein's insights, although they remained a mystery to most people, ensured that science would never be the same again.

EINSTEIN AND THE BOMB

Einstein's fourth paper of 1905 contained the famous equation $E = mc^2$, which shows the relationship between energy (E) and mass (m). The amount of energy locked up in a piece of matter is equal to its mass multiplied by the speed of light squared (c^2). As the speed of light is very fast, a mass can, potentially, be converted into a huge amount of energy. When Einstein wrote this equation, it was purely scientific. But as World War II approached, the Nazis began to use the principle to develop the atomic bomb. In 1939, Einstein and other scientists wrote to U.S. president Franklin D. Roosevelt, urging the United States to carry out similar research. However, when the United States finally dropped an atomic bomb on Hiroshima, Japan, in 1945, Einstein was appalled by the devastation. He became a committed pacifist.

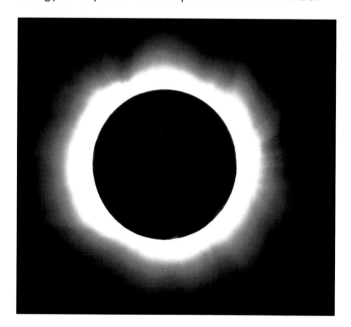

The light-bending effect that proves Einstein's Theory of Relativity can be measured during a total solar eclipse. This eclipse was seen in many parts of the world on August 11, 1999.

Hiroshima, Japan, after the atomic bomb was dropped. By the end of 1945, about 140,000 people had died as a result of the bomb's terrible power.

GRAVITY AT WORK

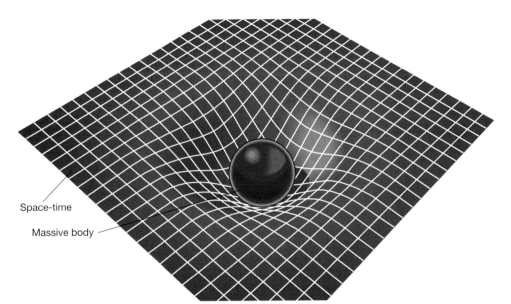

Einstein explained how gravity works by showing that massive objects cause the space around them to curve. As a result, less massive objects "fall" toward them.

Some years before he published the General Theory of Relativity, Einstein developed the idea that massive objects in space cause light to bend. In this letter of 1913, written in German to American astronomer George Ellery Hale, Einstein uses a diagram to show the bending effect. This effect was proved when there was a total eclipse of the sun in 1919. This event created conditions in which British astronomer Arthur Eddington was able to test whether the path of starlight was affected by the sun's massive bulk. It was, exactly as Einstein had predicted.

0.84" is the number of seconds (") by which Einstein predicts the starlight will be deflected. A second here is a subdivision of an angle.

Stern is German for star.

This line shows the expected path of the starlight.

Sonne is German for sun.

CHAPTER 4

GEOLOGISTS

CHARLES LYELL
AND UNIFORMITARIANISM

The holy books of many religions explain when and how the world was created. These ideas went unchallenged for thousands of years. Then some people began to take a more scientific approach, investigating the Earth's surface for evidence. By the nineteenth century, geology was an accepted science, and Charles Lyell (1797–1875) was one of its most important figures.

Many early geologists were miners seeking coal and iron to supply the factories of the Industrial Revolution. In the eighteenth century, Abraham Werner, a teacher at a German mining academy, proposed a geological theory. A flood, he said, had once covered the Earth, and the planet's rock strata were formed from minerals in the water that sank to the seabed or crystallized as the

Charles Lyell

ocean evaporated.

Scottish geologist James Hutton disagreed with Werner's ideas. In 1795, he published his own views in *Theory of the Earth*. The Earth's surface, Hutton said, had not been shaped by floods or other sudden catastrophes. It had developed—and was still developing—as a result of processes such as erosion, sedimentation, and heating. Heating occurred inside the Earth, making volcanoes erupt and pushing rocks up to create folds and mountains. New rocks formed as volcanic material under the surface solidified.

Hutton's theory, which he called uniformitarianism, won support from other geologists. They included his fellow Scot, Charles Lyell. As a young law student with a strong interest in geology, Lyell had traveled widely in England, France, and other parts of Europe during the early nineteenth century. On his trips, he had studied the landscape, examining any fossils he found there. These thorough investigations had convinced him that Hutton's ideas were correct.

Eventually, Lyell decided to write a book in support of Hutton's theory. This three-volume work is called *The Principles of Geology* (1830–33). In it, Lyell explains how physical processes

In this cartoon, James Hutton chips away at rocks in the shapes of the faces of his scientific enemies.

created the landscape. He also uses fossil evidence to date the events he describes. The work was an instant hit. Among its readers was naturalist Charles Darwin (see pages 46–9), who later used some of Lyell's ideas.

During the course of his life, Lyell made many more contributions to the study of the Earth, most notably developing an early form of geological time scale, as well as ensuring that uniformitarianism was widely accepted. By the late nineteenth century, geology had become a strong, evidence-based science with a sound framework on which future scientists could build.

Lyell formulated many of his theories by studying the landscape. He was especially intrigued by the volcanic plugs—lumps of solidified magma—in central France, seen here.

TELLING THE TIME

William Smith, a nineteenth-century British surveyor, examined land where people wanted to build canals. As he worked, Smith noticed animal fossils in the rock strata. He realized that certain types of fossils occurred in the lower, older layers of rock, while other types occurred in the younger layers on top. In other words, fossil types were a guide to the age of rocks. This insight was a great help to early geologists, but during the twentieth century, far more accurate dating techniques were devised.

The first of these two excerpts is taken from a letter Lyell wrote just before the publication of *The Principles of Geology*. The second comes from a letter written during a trip to the United States. It discusses one of the fossils Lyell examined and outlines the problems encountered by a geologist whose ideas conflicted with Biblical teaching.

Abstract is another word for summary.

Naples: January 15, 1829.

...My work is in part written, and all planned. It will not pretend to give even an **abstract** of all that is known in geology, but it will endeavor to establish the principle of reasoning in the science; and all my geology will come in as illustration of my views of those principles... which, as you know, are neither more nor less than that no causes whatever have from the earliest time to which we can look back, to the present, ever acted, but those now acting; and that they never acted with different degrees of energy from that which they now exert.

These few lines are a brief description of the theory of uniformitarianism.

Philadelphia: April 27, 1846.

I had some thoughts before my arrival here, to draw up a paper on what I learnt and observed... in the Pennsylvanian Coal Field... I have satisfied myself that Dr. King is right in believing that he has discovered in the middle of the coal formation, the foot tracks of a large reptilian quadruped or animal allied to the **Cheirotherium**... Dr. King was a beginner in geology when he first found, and to his credit appreciated duly, the importance of the Cheirotherium tracks. He is a man of thirty years of age, and in an extensive medical practice, who has suffered some persecution, professionally and socially, for believing the world to be more than 6,000 years old...

The **Cheirotherium** was a type of ancient, four-legged reptile.

In the seventeenth century, an archbishop named James Ussher used Bible texts and the works of ancient writers to calculate that the world was created in 4004 B.C.

ALFRED WEGENER
AND CONTINENTAL DRIFT

In the nineteenth century, geologists discovered how the Earth's rocks and natural features formed and changed. In the twentieth century, a few of them, including Alfred Lothar Wegener, began to investigate a puzzle on a larger scale. They suspected that the world's great land masses, the continents, had moved over time. But they needed evidence to prove their theory to fellow scientists.

Anyone looking at a world map can see that many separate pieces of land would fit neatly into one another if they were pushed together across the sea. For example, the point on the east coast of

South America is just the right shape to fit underneath the bulge on the west coast of Africa. In the nineteenth century, some geologists began to ponder what this might reveal about the history of the Earth.

Austrian scientist Eduard Suess, professor of geology at Vienna University in 1861, was the first to consider this question in detail. He discovered that rocks on coasts that faced one another across the sea often matched exactly. He also found that many lands now separated by ocean contained fossils of the same plants and animals. All this

ALFRED LOTHAR WEGENER
(1880–1930)

Alfred Lothar Wegener (below), born in Berlin, Germany, in 1880, studied astronomy at universities in Germany and Austria. His other great interest was meteorology, and in 1908 he became a lecturer in both subjects at the University of Marburg in Germany. While there in 1915, he wrote his famous book *Die Entstehung der Kontinente und Ozeane*. Four years later, after World War I, Wegener joined the meteorological department of the German Marine Observatory and began making trips to Greenland to carry out research. In 1924, he was appointed professor of meteorology and geophysics at Graz University in Austria. Wegener died during his fourth trip to Greenland.

CONTINENTAL DRIFT

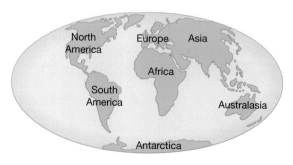

These modern continental drift maps show the Earth's land masses as they were 180 million years ago (top), 120 million years ago (middle), and today (bottom).

evidence led Suess to claim that Australia, India, Africa, and South America had originally formed one vast southern continent. In the 1890s, he christened this landmass Gondwanaland.

In the early twentieth century, Wegener proposed an even more startling theory about the continents. His 1915 book *Die Entstehung der Kontinente und Ozeane (The Origin of Continents and Oceans)* made the bold suggestion that all the world's continents had once been a single giant supercontinent. He called this landmass Pangaea, Greek for "All Earth." Like Suess, Wegener used rock and fossil evidence to support his idea. He also pointed out that the continental shelves, the underwater ledges that surround continents, would fit together even better than the coastlines. They matched like the pieces of a jigsaw puzzle. Wegener's most daring step was to propose a theory about how the continents had changed position.

Wegener rewrote parts of *Die Entstehung der Kontinente und Ozeane* (1915) several times in order to add information provided by Alexander Du Toit and others. The English translations, from which this and the excerpt on page 43 are taken, include all these new facts. Below, Wegener explains how rocks and their formations can provide evidence of his continental drift theory.

18 THE ORIGIN OF CONTINENTS AND OCEANS

Upper Carboniferous

Eocene

Lower Quaternary

FIG. 4. Reconstruction of the map of the world according to drift theory for three epochs.
Hatching denotes oceans, dotted areas are shallow seas; present-day outlines and rivers are given simply to aid identification. The map grid is arbitrary (present-day Africa as reference area; see Chapter 8).

These three maps from *Die Entstehung der Kontinente und Ozeane* show how Wegener believed the world's continents had begun as a single landmass (top), then gradually drifted apart.

A **rift** is a giant crack in the Earth's crust.

By comparing the geological structure of both sides of the Atlantic, we can provide a very clear-cut test of our theory that this ocean region is an enormously widened **rift** whose edges were once directly connected, or so nearly as makes no difference... alkali-rich rock is... to be found on exactly corresponding stretches of coastline: on the Brazilian side, at various places in the Serra do Mar... on the African side, on the coast of Lüderitzland, at Cape Cross north of Swakopmund, but also in Angola... both in Brazil and South Africa... beds yield the famous diamond finds. In both these regions the peculiar type of **stratification** known as "pipes" occurs. There are white diamonds in Brazil in Minas Geraes State and in South Africa north of the Orange River...

Stratification means rock layering.

Wegener called his theory continental drift. It suggested that Pangaea broke up about 200 million years ago and that the resulting pieces slowly drifted apart. On research trips to Greenland, Wegener had taken measurements that showed this drifting was still taking place. The measurements themselves were very inaccurate, and have since been revised using lasers, but the movement was evident. Many geologists ridiculed Wegener's idea. Their main objection was that the continents, which are largely made of granite, could not have moved through the solid basalt rock below.

After his book was published, Wegener continued his academic career. Meanwhile, two other geologists were carrying out research that supported his ideas. The first was South African Alexander Du Toit. While working for the Geological Commission of the Cape of Good Hope, Du Toit studied South African rocks in great detail. In 1923, he carried out similar work in South America. His results convinced him that the two continents had once been joined. Du Toit outlined his ideas in *A Geological Comparison of South America with South Africa* (1927).

Wegener's second supporter was British geologist Arthur Holmes. In 1929, Holmes proposed an idea that helped explain continental drift. Holmes's idea was that heat generated inside the Earth caused rocks below the surface to circulate in convection currents. As the rocks rose, Holmes thought, they pushed up and out with great force, driving continents apart.

Despite these developments, most geologists continued to reject the idea of continental drift. Its mechanism was not understood until years later.

ROCKS AND RADIOACTIVITY

Using fossils to date rock layers was a major breakthrough for geologists. But this method could reveal only the relative ages of rocks—that is, if a particular rock was older or younger than another. The absolute age of rocks—how many years they had existed—remained a mystery. Then, in the twentieth century, scientists discovered that the atomic nuclei of radioactive materials emit particles at a fixed rate. These particles are often trapped inside radioactive rocks. Once their emission rate is known, scientists can determine the age of a rock by calculating how many particles it contains. Arthur Holmes used radioactive techniques to devise the first geological time scale that divided the Earth's history into dated periods. It appeared in 1913.

Arthur Holmes

Ernest Rutherford made the following statement in 1907. It indicates he was the first person to realize that the rate of radioactive decay could be used to date rocks.

If the rate of production of helium from known weights of the different radio-elements were experimentally known, it would thus be possible to determine the interval required for the production of the amount of helium observed in radioactive minerals, or, in other words, to determine the age of the mineral.

Alpha particles, a type of radioactivity, are chemically the same as atomic nuclei of the element helium.

FOSSIL EVIDENCE

The location of plant and animal fossils provided important evidence for continental drift theory. Fossils of the *Glossopteris* fern, found in Antarctica, Australia, India, Africa, and South America, supported the idea of an ancient southern supercontinent. Fossils of an extinct reptile named *Mesosaurus* were discovered in South Africa, Brazil, Paraguay, and Uruguay, again indicating that Africa and South America were once joined. The link between Europe and North America was suggested by the presence of *Diadectid* insect fossils in both places.

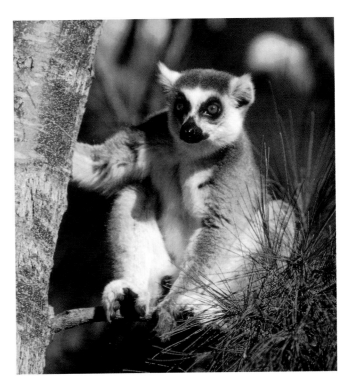

A ring-tailed lemur from Madagascar, an island off the east coast of Africa. All twenty-one species of lemurs are native only to that island. Madagascar split from Africa about sixty million years ago, so the animals there evolved separately.

The locations of living plants and animals, as well as fossils, can help geologists figure out where continents joined and when they separated. In this second excerpt from his book, Wegener explains what Australia's animals reveal about the Earth's past. He has already stated that they can be divided into three groups, and that each group has lived in the country for a different length of time.

Fauna means animal life.

Correlation means relationship or connection.

Marsupials are mammals whose young are born in an undeveloped state and continue to grow in a pouch outside their mother's body. They include kangaroos, koalas, and opossums.

The **Sundas** are a chain of islands in the Pacific and Indian oceans. They include Java and Borneo.

The most ancient element, generally encountered in southwestern Australia, shows some interrelationship with the fauna of India... Here warmth-loving animals are representative of the relationship, and so are the earthworms... The correlation dates back to the time when Australia was still joined to India. The second Australian faunal element... includes those peculiar mammals, the marsupials... This element shows points of relationship with South American fauna. Apart from Australia... marsupials now live mainly in South America... The third Australian faunal element is the most recent, and emigrated from the Sundas; it has... already taken over the northeast of Australia. The dingo (wild dog), rodents, bats and other mammals immigrated to Australia in the post-Pleistocene... This threefold subdivision... agrees most elegantly with drift theory. One need only look at the three reconstruction maps [see page 41] and the explanation will be immediately forthcoming.

The **Pleistocene** is a geological era that began about 1.8 million years ago and ended about 10,000 years ago.

HARRY HESS
AND SEA-FLOOR SPREADING

Geologists eventually solved the puzzle of continental drift by examining the Earth's ocean floors. American geologist Harry Hammond Hess played a major role in this investigative work.

Exploration of the world's ocean floors began in earnest during the nineteenth century. Gradually, using sonar and other techniques, scientists discovered that sea floors were not completely flat but made up of mountains, trenches, and other natural features. Then, in the 1950s, American oceanographer William Maurice Ewing showed that a 40,000-mile (65,000-km) long mountain range snaked around the world in the middle of the ocean floor. In 1957, he also demonstrated that giant rifts (cracks) ran all along the center of this mid-ocean ridge.

Harry Hess, professor of geology at Princeton University, followed these developments closely. He had read Wegener and Du Toit's works and was intrigued by the continental drift theory. He doubted its truth, however, because he knew continents did not have the energy to move across the Earth. Then Ewing's findings, and his own discovery that ocean floor rocks are younger than the rocks of the surrounding continents, gave him a revolutionary idea.

In 1962, Hess revealed his idea in a paper called "History of Ocean Basins." It was a description of the way he believed the ocean floor had formed. Molten rock called magma, Hess suggested, was always rising from inside the Earth and oozing through the rifts in the mid-ocean ridge. As the magma oozed, it pushed existing rock away from the ridge, then solidified to form new, younger rock. Hess's theory became known as sea-floor spreading.

British geologists Fred Vine and Drummond Matthews set out to test Hess's idea. They knew that the Earth's magnetic field periodically reverses itself, the north magnetic pole becoming the south, and the south becoming the north. They also knew that, when magma cools, evidence of the field's direction at the time is preserved in the solid rock that forms. Gradually, series of rock "stripes" with opposing magnetic fields are created. If Hess was right, the magma from the mid-ocean rifts would have formed stripes in the surrounding ridge.

Vine and Matthews found that the ridge was indeed made up of rock stripes, which could be dated. In addition, the stripes were symmetrical on the two sides of each central rift, indicating that both had formed from magma leaving the rift at the same time. Vine and Matthews published their findings in 1963, and sea-floor spreading was accepted. Scientists soon

SEA-FLOOR SPREADING

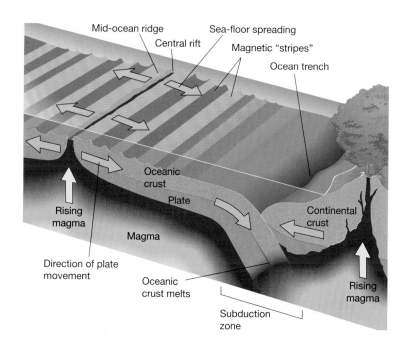

Mid-ocean ridge
Central rift
Sea-floor spreading
Magnetic "stripes"
Ocean trench
Oceanic crust
Plate
Continental crust
Rising magma
Magma
Direction of plate movement
Oceanic crust melts
Subduction zone
Rising magma

As magma rises through rifts in the sea floor, it forces the surrounding rock apart. As it solidifies, it creates new rock.

realized that Hess's theory helped explain continental drift. The continents did not, in fact, drift apart. They were pushed apart by rock emerging from the ocean floor.

HARRY HAMMOND HESS (1906–69)

Harry Hammond Hess (below) was born in New York City. After graduating from Yale University in 1927, he worked as a geologist in Northern Rhodesia. In 1934, Hess joined the staff at Princeton University. As he pursued an academic career, Hess served in the Navy. While surveying the Pacific Ocean floor, he discovered many flat-topped mountains, previously unknown features that he called guyots. His survey work prompted Hess's interest in submarine geology and led him to develop the sea-floor spreading theory.

PLATE TECTONICS

New ideas about the Earth's structure improved geologists' understanding of sea-floor spreading. The Earth's crust—the rock that makes up the ocean floor and the continents—forms only the top part of a deeper layer. In the 1960s, Canadian geologist John Tuzo Wilson and others showed that this layer is divided into about fifteen giant pieces called "plates." The rifts in the ocean floor are plate edges. As new rock is formed, it forces the plates apart, and the plates carry the continents with them. The opposite process occurs in subduction zones, where plate edges collide and one forces the other below the surface. The study of plate movement is known as plate tectonics.

The San Andreas Fault, in California. It is a strike-slip fault, produced when two plates slide past each other.

Hess was delighted that his sea-floor spreading theory caused such a flurry of interest. But he was alarmed that so much of the research that followed was not carefully planned. In the introduction to a 1965 paper, he includes this warning.

A **hypothesis** is an untested scientific theory. Hess is saying that it is pointless for scientists to carry out research and gather information if they have no idea what they are trying to prove or disprove. A hypothesis, even if it is wrong, still provides the necessary framework for worthwhile scientific research.

To bring the problem into focus and guide continued exploration, coordinating **hypotheses** are needed and necessary. Even incorrect or partly incorrect speculations serve to identify the crucial observations needed for progress. Blind... collection of data without a framework of hypothesis by which it can be tested is wasteful and commonly unproductive, and leads to an accumulation of an indigestible mass of data of minor significance.

CHAPTER 5

BIOLOGISTS
CHARLES DARWIN
AND EVOLUTION

As a young man, British naturalist Charles Robert Darwin briefly trained for the Anglican priesthood. Eventually, however, he abandoned his religious studies for a scientific career. The results were profound, producing not only a revolution in biology, but also turmoil in the church.

In 1831, following his graduation from Cambridge University, Darwin was appointed naturalist on board the HMS *Beagle*, a ship heading for South America and the Pacific Ocean to carry out coastal surveys. The *Beagle* remained at sea for five years. During that time, Darwin explored many places, marveling at the variety of animals and plants and collecting specimens. He also examined landscapes,

HMS *Beagle*, the ship on which Darwin made his momentous journey. It carried small boats that he and other passengers used for exploring small islands and rivers.

using *The Principles of Geology* by Charles Lyell as his guide.

In 1839, three years after his return to England, Darwin

published an account of his trip, *A Naturalist's Voyage on the* Beagle. But he continued to ponder the deeper significance of what he

---

CHARLES DARWIN (1809–82)

Charles Darwin (right) was born in Shrewsbury, England. He enrolled to study medicine at Edinburgh University in 1825, but soon realized he had made a mistake. The lectures bored him, and the demonstrations on patients were an ordeal. He left for Cambridge University, where he prepared for a life as a priest in the Church of England. Soon, however, he grew far more interested in geology and botany. Darwin became friends with Professor John Henslow, who recommended Darwin for a job on the *Beagle*. In addition to an eventful career, Darwin had a full personal life—he married his cousin and they had ten children. Eventually, he became ill with Chagas disease, which was caused by a parasite he had picked up during his voyage to South America. He died in 1882 and was buried in Westminster Abbey.

had seen. Darwin had been astounded by the sheer number of animal species, many of which differed only in small details. For example, on the Galapagos Islands he had counted fourteen types of finches. They were all similar, but their beaks varied according to the type of food they ate. In the same way, Galapagos tortoises had necks of different lengths, depending on whether they ate leaves from trees or ground plants.

All this evidence led Darwin to develop a new theory. He began to doubt the Christian teaching that God had created each animal species as unchanging. Instead he began to think that each species had evolved, or changed, over time so that it could survive. In the case of the Galapagos finches, Darwin suggested that a single finch species had arrived on the islands from mainland South America. It had then evolved into many species, each adapted to conditions and food sources in the particular area where it settled.

Now Darwin began to consider what the ➲

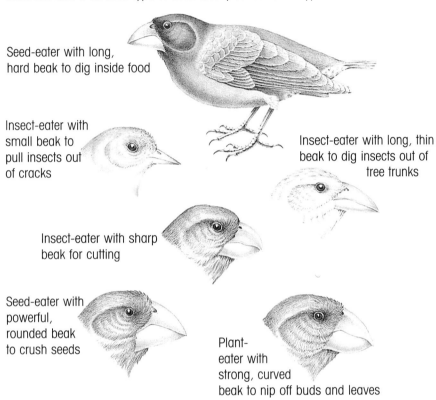

Six of the fourteen kinds of finches that Darwin found on the Galapagos Islands. Each one has a different type of beak ideally suited to the type of food it eats.

Seed-eater with long, hard beak to dig inside food

Insect-eater with small beak to pull insects out of cracks

Insect-eater with long, thin beak to dig insects out of tree trunks

Insect-eater with sharp beak for cutting

Seed-eater with powerful, rounded beak to crush seeds

Plant-eater with strong, curved beak to nip off buds and leaves

The following excerpt comes from the third chapter of *The Origin of Species,* "The Struggle for Existence." The words provide a vivid description of this struggle, which leads to the "survival of the fittest."

When we look at the plants and bushes clothing an entangled bank, we are tempted to attribute their proportional numbers and kinds to what we call chance. How false a view this is!... [It] has been observed that the trees now growing on the ancient Indian mounds, in the Southern United States, display the same beautiful diversity and proportion of kinds as in the surrounding virgin forests. What a struggle between the several kinds of trees must here have gone on during long centuries, each annually scattering its seeds by the thousand; what war between insect and insect—between insects, snails, and other animals with birds and beasts of prey—all striving to increase, and all feeding on each other or on the trees or their seeds and seedlings, or on the other plants which first clothed the ground and thus checked the growth of the trees!

Early indigenous cultures that flourished in the southeastern United States, such as the Adena and Hopewell nations, built mounds that were often used for burials or religious rituals.

mechanism for evolution might be—that is, how species changed. The answer came to him after he read an essay by economist Thomas Malthus. It said that when human populations could not produce enough food for all, the weakest people starved, died of disease, or were killed in fighting. In other words, only the strong survived. Darwin saw that this could apply to animals, too.

Darwin began to work out the details. He realized that when animals were competing for food, or struggling to live in a harsh environment, their physical characteristics affected their ability to survive. For example, birds with strong beaks can crack open seeds easily, and animals with powerful legs can escape their predators. So, Darwin reasoned, creatures with useful characteristics lived long enough to breed and pass them on to offspring. By contrast, animals without these characteristics died out. Darwin called this process natural selection.

Darwin agonized for many years about revealing his theories to the public, because he foresaw the uproar they would cause. Eventually, in 1859, he published *The Origin of Species by Means of Natural Selection*. Although the book became a best-seller, it provoked fury among Christians who believed in the literal truth of the Bible. The idea that natural selection could lead to the changing of one species into another caused particular anger. This was

because it implied humans were not created directly by God but had evolved like other creatures. The book only hinted at this conclusion, but controversy still raged. Despite the hostile reaction, Darwin continued his research and published several more books. They included *The Descent of Man* (1871), which discussed human evolution in some detail. Darwin's works are still widely read today, and his ideas are generally accepted by the scientific community.

ALFRED RUSSEL WALLACE

Darwin was not the only person to formulate the theories of natural selection and evolution—Welshman Alfred Russel Wallace reached similar conclusions. In the first half of the nineteenth century, Wallace traveled widely across South America and the Malay Archipelago. Like Darwin, he collected specimens, wrote books, and thought deeply about how species might have developed. In 1858, Wallace wrote a paper called "On the Tendency of Varieties to Depart Indefinitely from the Original Type" and sent it to Darwin for comment. Darwin realized Wallace had the same ideas as his, so he quickly finished *The Origin of Species* before Wallace had the chance to publish.

"DARWIN'S BULLDOG"

Darwin knew *The Origin of Species* would cause a great uproar, but he was a shy man who had no desire to join in the public debate. Fortunately, he had many able defenders who were willing to do just that. Chief among them was British biologist Thomas Henry Huxley (below). In 1860, Huxley faced two of Darwin's most outspoken opponents at a specially arranged meeting in Oxford, England. They were Samuel Wilberforce and a fossil expert named Richard Owen. Despite his reputation as a smooth speaker, "Soapy Sam" Wilberforce was outwitted by Huxley's firm grasp of the arguments and refusal to be angered. As a result of this and many other displays of debating skill, Huxley became known as "Darwin's bulldog."

This extract comes from Chapter Four of *The Origin of Species*, "Natural Selection." Darwin explains how this process caused some insects and birds to develop camouflage colorings, making them difficult for predators to see and catch.

Although natural selection can act only through and for the good of each being, yet characters and structures, which we are apt to consider of trifling importance, may thus be acted on. When we see leaf-eating insects green, and bark-feeders mottled grey; the alpine ptarmigan white in winter, the red-grouse the color of heather, and the black-grouse that of peaty earth, we must believe that these tints are of service to these birds and insects in preserving them from danger. Grouse... are known to suffer largely from birds of prey; and hawks are guided by eyesight to their prey... Hence I can see no reason to doubt that natural selection might be most effective in giving the proper color to each kind of grouse...

ON

THE ORIGIN OF SPECIES

BY MEANS OF NATURAL SELECTION,

OR THE

PRESERVATION OF FAVOURED RACES IN THE STRUGGLE FOR LIFE.

By CHARLES DARWIN, M.A.,

FELLOW OF THE ROYAL, GEOLOGICAL, LINNÆAN, ETC., SOCIETIES;
AUTHOR OF 'JOURNAL OF RESEARCHES DURING H. M. S. BEAGLE'S VOYAGE
ROUND THE WORLD.'

LONDON:
JOHN MURRAY, ALBEMARLE STREET.
1859.

The right of Translation is reserved.

The title page of Darwin's revolutionary book

This red grouse (right) is barely visible among the heather. Darwin believed the bird's camouflage was a product of evolution designed to help it hide from predators.

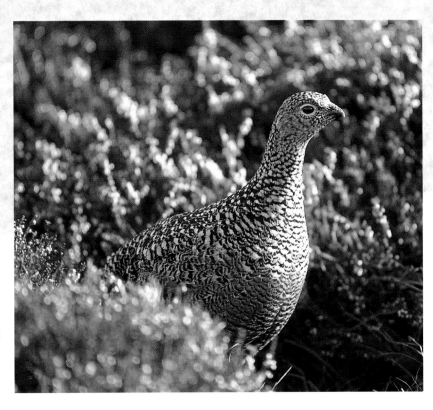

Many people mocked Darwin's theory of natural selection. Benjamin Disraeli, who became British prime minister in 1868, was one of the theory's detractors. The words he uttered on the subject, quoted here, have since become famous.

The question now placed before society is this, "Is man an ape or an angel?" I am on the side of the angels.

GREGOR MENDEL
AND GENETICS

Charles Darwin believed the offspring of any two creatures was a half-and-half mixture of its parents. Yet he also knew that if this inheritance pattern, known as blending, were true, natural selection would not work. An adult would pass on just half of a useful characteristic to its young, which in turn would pass on only a quarter. Darwin never solved this puzzle, but Austrian monk Gregor Johann Mendel did.

In 1856, Mendel began conducting plant-breeding experiments in the garden of his monastery. His goal was to show how different plant characteristics were passed from generation to generation. He decided the best way to do this would be to run several separate cross-breeding experiments, each tracking the path of a single characteristic. Mendel chose pea plants for his experiments. The seven characteristics he selected to study included height, pod color, and pea shape.

Mendel's first experiments involved cross-breeding tall and short pea plants. To his surprise, no plants of intermediate height were produced—they were all tall. This result immediately disproved the blending theory. However, when the new tall plants were cross-bred, one short plant was produced for about every three tall plants. This ratio also occurred in

MENDEL'S PEA PLANT EXPERIMENT

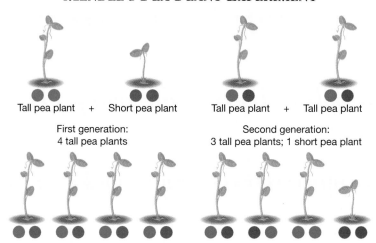

Tall pea plant + Short pea plant   Tall pea plant + Tall pea plant

First generation: Second generation:
4 tall pea plants   3 tall pea plants; 1 short pea plant

This diagram of Mendel's pioneering experiment shows how the genes for tallness and shortness were passed through two generations of plants.

experiments for several other characteristics. Mendel continued this work for eight years, painstakingly recording his results. Then he came to some important conclusions.

Inheritance, Mendel said, was governed by "factors." Each parent plant contributed a single factor for each

GREGOR MENDEL (1822–84)

Gregor Johann Mendel (below) was born in the Austrian town of Heinzendorf (now Hynčice in the Czech Republic). His father was a farmer, and as Mendel worked with him in the fields, he developed a great interest in plants. After studying at Olmütz University, Mendel became an Augustinian monk. Starting in 1843, his home was a monastery that was already a center of scientific research. There, between 1856 and 1863, Mendel carried out his famous plant experiments. In 1868, Mendel became abbot of his monastery, and afterward had little time for scientific study.

characteristic to its offspring, so that it had a total of two. This combination determined which form of a characteristic the offspring would inherit. If the two factors were the same—if both were for tallness, for example—then the plant would inherit that trait. But if the two were different, then one factor—the dominant factor—would cancel out the other, recessive factor. This theory explained the results of the pea plant experiments.

In 1865, Mendel read his findings to the Brünn Natural History Society and published them in its journal. Sadly, no one took much notice. It was only after Mendel's death in 1884 that their significance was revealed. "Factors" were what scientists now call genes, and Mendel had founded the science of genetics.

The excerpt below is taken from *Experiments in Plant Hybridization* (cross-breeding), the paper Mendel read to the Brünn Natural History Society in 1865. In it, Mendel lists the seven pea plant characteristics that he examined.

AMENDING MENDEL

Hugo de Vries, professor of biology at Amsterdam University, rediscovered Mendel's work in 1900. Among the many scientists who read it shortly afterward was Thomas Hunt Morgan (right), professor of experimental zoology at Columbia University in New York City. He doubted Mendel's theories until 1908, when he began to breed fruit flies. The results of his own experiments proved Mendel was mostly correct. However, Morgan made one vital amendment to the monk's laws. Mendel had stated that each characteristic is passed by genes from parent to offspring independently. Morgan showed that groups of genes make up larger structures called chromosomes, and that genes on the same chromosome are often inherited together.

Mendel used this word to mean the endosperm, the food material inside seeds that surrounds the growing embryo.

The characters which were selected for experiment relate:
1. To the difference in the form of the ripe seeds...
2. To the difference in the color of the seed albumen...
3. To the difference in the color of the seed-coat...
4. To the difference in the form of the ripe pods...
5. To the difference in the color of the unripe pods...
6. To the difference in the position of the flowers...
7. To the difference in the length of the stem...

The first, handwritten page of the paper that Mendel read to the Brünn Natural History Society

LOUIS PASTEUR
AND GERM THEORY

Frenchman Louis Pasteur (1822–95) was both a chemist and a microbiologist—a scientist who studies organisms so small they can only be seen under a microscope. Pasteur's major breakthrough was his discovery that microorganisms are responsible for many diseases. Armed with this knowledge, he developed vaccines to prevent diseases from spreading.

Pasteur became professor of chemistry at Lille University in France in 1854. The city of Lille was an important beer-brewing center. One of the beer-making processes involved allowing sugar beet to ferment so that it turned into alcohol. At the time of Pasteur's arrival, sugar beet in many of the vats was turning sour instead of forming alcohol. Pasteur was asked to find out why. He examined both the alcoholic and the sour liquids. In the alcoholic liquid, he saw specks of yeast, a type of microorganism. He figured out that the specks caused fermentation by breaking down sugar in the liquid to produce alcohol and carbon dioxide. No one had understood this process before. Pasteur found a different microorganism in the sour liquid. This one was causing fermentation, too, but the product was lactic acid, not alcohol. Pasteur told the brewers to destroy all the liquid contaminated by these microorganisms. They did so, and the problem was solved.

Pasteur now wanted to discover where microorganisms came from. Were they in the air, as he believed, or spontaneously produced by rotting substances, as other scientists claimed? To find out, he did an experiment. First he put some sterilized fluid into two groups of swan-necked flasks. Then he snapped the necks off one group of flasks, allowing dust to enter. The fluid in these flasks was soon full of microorganisms, but the fluid in the other flasks remained clean. The "germs," as Pasteur now called the microorganisms, were in the air.

Pasteur put his new understanding to practical use by inventing pasteurization treatment—the heating of wine, milk, and other liquids to high temperatures in order to kill harmful microorganisms. By this time he was convinced that such germs were the cause of many human and animal diseases, too. Soon he discovered the germ that caused the highly infectious disease anthrax, and in 1882 he created an effective vaccine from a weakened form of this

A replica of one of the swan-necked flasks that Louis Pasteur used in his germ theory experiments. It is filled with a nutrient-rich broth similar to the jelly Pasteur describes in the document on page 53.

Louis Pasteur. An immensely dedicated man, he suffered two serious strokes in 1868 and 1887 but continued working until his death.

microorganism. In 1885, he produced a successful vaccine against rabies and used it to save the life of a young boy.

Pasteur became a public hero. In 1888, a medical research institute named after him was opened in Paris. Today scientists at the Pasteur Institute still seek cures for diseases such as AIDS. If it were not for the discovery that many diseases are caused by germs, they would have little hope of success.

SAVING THE SILKWORMS

In 1865, Pasteur was called away from Paris to southern France, where the silk industry was in crisis. Silkworms were dying in the thousands, so they could not spin the shining threads on which the industry depended. Worm breeders had a name for the main disease causing all the damage—pebrine—but had no idea how it was caused. Working with his usual thoroughness, Pasteur was able to identify the responsible microorganism. Before returning to Paris, Pasteur explained to the breeders how to make their industry disease-free.

The Pasteur Institute in Paris today. Its founder is buried in a specially built crypt underneath.

Many nineteenth-century scientists disagreed with Pasteur's view that disease-causing microorganisms were present in the air. These scientists believed in "spontaneous generation," the idea that diseased or decaying substances produced the microorganisms as they broke down. But Pasteur was determined. In this excerpt from a lecture he gave at the Sorbonne, he explains how he tried to grow microorganisms in a drop of water. Then he states why it did not work.

In all the immensity of creation, I have taken my drop of water and I have taken it full of rich jelly—that is, to use the language of science, full of elements most suited to the development of small beings. And I wait, I observe, I question it, I beg it to be so kind as to begin over again, just to please me, the primitive act of creation; it would be so fair a sight! But it is mute! It has been mute for years. Ah! That is because I have kept far from it, and still keep far from it, the only thing that it has not been given to man to produce. I have kept from it the germs that float in the air; I have kept from it life, for life is a germ and a germ is life. Never will the belief in spontaneous generation arise from the mortal blow that this simple experiment has given it.

ALEXANDER FLEMING
AND PENICILLIN

A petri dish prepared by Alexander Fleming to demonstrate the action of penicillin. The mold is inhibiting the growth of some of the bacteria streaks on the jelly, just as it did in the original dish in 1928.

Pasteur's germ theory was finally accepted, giving scientists a much better understanding of disease. However, many types of germ remained a threat to human and animal life. Scottish bacteriologist Alexander Fleming made a discovery that was to provide doctors with a powerful new way of overcoming germs.

In 1928, Fleming was appointed professor of bacteriology at St. Mary's Hospital in London. In the same year, he made his most important discovery. As part of his research into the treatment of influenza ("flu"), Fleming had to grow bacteria on jelly in special glass dishes. One morning he noticed that something had destroyed the bacteria on certain areas of jelly in an uncovered dish. Tests showed that this was a substance produced by a mold whose Latin name was *Penicillium notatum.* Fleming called it "penicillin."

Fleming knew that penicillin was effective against staphylococcus, the type of bacteria on the dish. Now he began experiments to discover which other types of bacteria it would destroy. The results were encouraging, but Fleming was unable to isolate the substance that killed the bacteria. Interest among

other scientists was low, and penicillin research gradually stopped.

In 1939, two other scientists, Australian Howard Florey and German-born Ernst Chain, returned to the subject. As World War II began, the British government encouraged their research, hoping it might lead to new battlefield treatments. By 1941, the scientists had produced a pure form of penicillin that worked as a drug. Wartime Britain could not afford to make it, however, so Florey turned to the United States for help. Mass production of penicillin

ALEXANDER FLEMING (1881–1955)

Alexander Fleming was born in Lochfield, Scotland. When he was sixteen, he left school and took a job in London as a shipping clerk. He wanted to pursue a medical career, so he began a course of study at St. Mary's Hospital Medical School. He graduated in 1908 and stayed at the school, leaving only temporarily to serve in World War I. When he returned, he devoted himself to the study of bacteria. In 1922, he discovered lysozyme, an enzyme present in tears, saliva, and other body fluids that is an effective antiseptic. Six years later, while professor of bacteriology at St. Mary's, Fleming made the discovery for which he is now famous.

Fleming in his laboratory at St. Mary's Hospital, Paddington, in 1909

began in the United States in 1943, and a few years later this new antibiotic became widely available.

Fleming's work ushered in a new era of medicine. Soon doctors were using penicillin and other antibiotics to treat many diseases. But now, in the twenty-first century, they are more wary. Overuse of antibiotics has enabled some bacteria to develop resistance to them. Today the search is on for new types of these life-saving drugs.

FLOREY AND CHAIN

Howard Florey and Ernst Chain came from opposite sides of the world. Florey was born in Adelaide, Australia, in 1898 and studied there before winning a scholarship to Oxford University in England. He returned to Oxford in 1935 as head of the Sir William Dunn School of Pathology. Chain was born in Berlin, Germany, in 1906. In 1933, when the Nazis came to power, he left for England, enrolling at Oxford University in the same year as Florey. Ten years later, Florey, Chain, and Fleming shared the Nobel Prize for their work on penicillin.

The Nobel medal awarded to Florey, Chain, and Fleming in 1945

By the mid-1940s, doctors were prescribing penicillin. It was still a new drug, so they needed guidance on how to use it. To help them, one publisher released a book called *Penicillin: Its Practical Application* (1946). The book was edited by Fleming, who also wrote the introduction. In this excerpt from the introduction, he describes one of the methods he used to test penicillin.

Petri dishes are shallow, round glass dishes with lids. They are used in laboratories for growing microorganisms.

Microbes are bacteria.

Incubation is the period during which the bacteria were left to grow.

Gutter here means groove.

With a knife a gutter was cut out of the agar in a Petri dish and was filled with agar containing... the mold. Different microbes were then planted in streaks from the edge of the plate to the gutter. After incubation some of these were found to grow right up to the gutter while others were inhibited for a considerable distance. This showed that the mold... had a powerful inhibitory action on some bacteria but not on others, and that among the sensitive bacteria were some of the commonest agents of infection in man.

Agar is a type of jelly on which bacteria are grown.

Inhibited here means held back or prevented from growing.

FIG. 7.—Gutter method of demonstrating selective inhibition

PENICILLIN

1.	*Esch. coli*
2.	Staphylococcus
3.	Haemolytic streptococcus
4.	Gonococcus
5.	*B. diphtheriae*
6.	*B. anthracis*
7.	*B. typhosus*

LYSOZYME

1.	*Esch. coli*
2.	Staphylococcus
3.	Sarcina
4.	*M. lysodiekticus*
5.	*B. subtilis*
6.	Streptococcus
7.	*B. typhosus*

This diagram from the book shows how the penicillin in the gutter (image on left) stopped bacteria of types 2 to 6 from growing.

CRICK AND WATSON AND DNA

By the 1950s, thanks to Mendel, Morgan, and others, scientists had a basic understanding of the role genes and chromosomes played in passing characteristics from one generation to the next. They also knew the chemical composition of these vital pieces of matter. The structure of genes remained a mystery until the team of Francis Crick and James Watson made a major breakthrough.

In 1869, Swiss biochemist Friedrich Miescher had discovered a substance that seemed to occur in all cell nuclei. He called it nuclein, which later became known as nucleic acid. There were at least two types. Other scientists then began to figure out what one particular type of nucleic acid contained. This turned out to be four chemical compounds of a kind known as bases, another type of chemical called a phosphate, and a sugar called deoxyribose. To distinguish it from other types, the acid was given its own special name of deoxyribonucleic acid—DNA.

Next, scientists turned their attention to two more puzzles. What was the exact location of DNA in cells, and what was its purpose? Both questions were answered by the 1940s. German chemist Robert Feulgen showed that DNA was found in chromosomes and the genes they contained. Canadian chemist Oswald Avery and his

FRANCIS CRICK (1916–) AND JAMES WATSON (1928–)

Francis Harry Compton Crick was born in Northampton, England, and James Dewey Watson was born in Chicago. Crick studied physics at University College in London and at Cambridge University. During World War II, he was employed by the Admiralty to design mines. Watson studied at the University of Chicago and the University of Indiana before moving to Copenhagen University and then to the Medical Research Council Unit at Cambridge. It was here that he met Crick. After their famous collaboration, Crick remained at Cambridge until 1977, when he joined the Salk Institute in San Diego, California. There he devoted himself to studying the brain and the nature of consciousness. In 1953, after the research on DNA structure was complete, Watson returned to the United States, eventually taking a job at the Cold Spring Harbor Laboratory in New York. From 1988 to 1992, Watson took a leading role in the groundbreaking Human Genome Project.

James Watson (left) and Francis Crick with their model of part of a DNA molecule in 1953

A DNA STRAND

Adenine
Thymine
Guanine
Cytosine

The four bases in DNA—adenine, thymine, guanine, and cytosine—make up the "rungs" of its twisted ladder shape. The sides of the ladder consist of alternating strips of phosphate (P) and deoxyribose sugar (S).

team showed that DNA from one kind of bacteria altered the nature of a second kind in which it was placed. In other words, DNA carried the hereditary material that produced the characteristics of living things.

Many scientists were now eager to learn more about DNA. Two of these were Englishman Francis Crick and American James Watson. By the 1950s, both were working in the Cavendish Laboratory at Cambridge University. Watson persuaded Crick to join him in the task of discovering the structure of DNA, and the two men set to work. Many other scientists had already carried out important research on the subject. Crick and Watson relied on X-ray diffraction pictures of DNA provided by Maurice Wilkins and Rosalind Franklin. It was the thought processes of Crick and Watson that finally revealed the shape of the DNA molecule: a graceful double helix that looked like a twisted ladder.

The significance of Crick and Watson's work was tremendous. Knowledge of DNA's structure enabled scientists to figure out exactly how genetic messages are passed between generations, leading to a deeper understanding of heredity and evolution. It also led to the development of genetic engineering —the alteration of genes to produce improved types of plants, animals, bacteria, vaccines, and more.

ROSALIND FRANKLIN

The role of Rosalind Franklin (below) in discovering the structure of DNA is much debated. Franklin was born in London and educated at Cambridge University. She joined King's College, part of London University, in 1951. While there, she was asked to help with the DNA project. Unfortunately, there was a personality clash between Franklin and Maurice Wilkins, who was also working on the DNA puzzle. Franklin felt her scientific contribution was underestimated because she was a woman. Most science historians now accept that Franklin's X-ray diffraction photographs provided Crick and Watson with some of the most important evidence for their discovery. There is disagreement, however, about whether she would have figured out the structure of DNA herself.

In 1988, Crick published _What Mad Pursuit_, a book that records his part in the DNA story and

describes his later research. In this excerpt, he spells out what made the DNA project like any other piece of scientific research—and what made it different.

I think what needs to be emphasized about the discovery of the double helix is that the path to it was, scientifically speaking, fairly commonplace. What was important was not the way it was discovered but the object discovered—the structure of DNA itself. You can see this by comparing it with almost any other scientific discovery. Misleading data, false ideas, problems of personal interrelationships occur in much if not all scientific work. Consider, for example, the discovery of the basic structure of collagen, the major protein of tendons, cartilage, and other tissues. The basic fiber of collagen is made of three long chains wound around one another. Its discovery had all the elements that surrounded the discovery of the double helix... Yet nobody has written even one book about the race for the triple helix. This is surely because, in a very real sense, collagen is not as important a molecule as DNA... It is the molecule that has the glamour, not the scientists.

CONCLUSION

Scientists have made enormous progress over the last two centuries. Their practical skills and theoretical brilliance have led to the discovery of everything from tiny subatomic particles and microorganisms to the powerful mechanisms that drive the Earth and the stars.

Such results did not come without effort. Many of the scientists featured in this book, such as Marie Curie, spent years carrying out painstaking experiments. Others, such as Charles Lyell, had to scour the Earth for evidence to support their theories. Albert Einstein sometimes became ill as he wrestled with complex mathematical problems. Despite all this hard work, setbacks were common. When breakthroughs were made, they were often mocked, as in the case of Alfred Wegener. Occasionally they caused public outrage, as Charles Darwin discovered.

Of course, the problems of research did not, in the end, deter the great scientists of the last two hundred years. Sheer curiosity, as well as the support of individual fellow scientists—and sometimes of the worldwide scientific community—spurred them on. Almost all scientists emphasize the great pleasure their discoveries gave them. Louis Pasteur said, "The reward is one of the keenest joys

Pure science has led to many beneficial technological advances. For example, the discovery of X rays made possible the development of CT (computerized tomography) scanners like this one. They use X rays to build up three-dimensional pictures of the body as an aid to medical diagnosis.

of which the human soul is capable." The rewards were often all the greater if the discovery had taken a long time. Public recognition in the form of a Nobel Prize was another boost, even if it came long after the work was complete.

In the twenty-first century, the role of the scientist is likely to be greater than ever. Discoveries in the

In the year 2000, scientists completed their map of the human genome—all the genes that humans contain. Now they know the exact location of each of our chromosomes, and are currently learning what each one does. The DNA of every person's genes is different, producing a "fingerprint"—a unique pattern of black bands (left).

fields of genetic engineering, artificial intelligence, and communications will almost certainly transform the future, just as discoveries relating to electricity, atomic structure, and evolution transformed the past. How this will happen, and what else might happen, is impossible to predict.

Recent centuries have provided ample proof that scientific discoveries can be put to destructive as well as good use. Albert Einstein's realization that energy is locked up in matter eventually led to the creation of the atomic bomb. In many people's view, this is the most obvious example of science gone astray. Robert Oppenheimer was director of the laboratory in Los Alamos, New Mexico, that made the bomb. Below he explains how tempting it is to continue research without thinking clearly about the consequences.

When an atomic bomb is dropped, a distinctive mushroom cloud appears over the site. This cloud appeared after a nuclear test in the desert of Nevada.

... it is my judgment in these things that when you see something that is technically sweet you go ahead and do it and you argue about what to do about it only after you have had your technical success. That is the way it was with the atomic bomb. I do not think anybody opposed making it; there were some debates about what to do with it after it was made.

Louis Pasteur recognized that science could be used for harmful purposes. But he knew from his own achievements that it could also be a force for good. In a speech he wrote for his seventieth birthday celebration, he stated his belief that good would triumph.

Gentlemen, you bring me the greatest happiness that can be experienced by a man whose invincible belief is that science and peace will triumph over ignorance and war... Have faith that in the long run... the future will belong not to the conquerors but to the saviors of mankind.

GLOSSARY

absolute magnitude The brightness of a star at a standard distance of 32.6 light years from the Earth.

alpha particle A type of particle made up of two protons and two neutrons. Alpha particles are emitted from atomic nuclei during some types of radioactive decay.

apparent magnitude The brightness of a star as it appears from Earth.

antibiotic Any of a class of drugs used to reduce or stop the growth of disease-causing bacteria.

artificial intelligence (AI) Advanced computer systems and programs that try to mimic human thought processes.

astrophysics The branch of astronomy that deals with the chemical composition, origin, and evolution of stars and other heavenly bodies.

atoms The smallest particles of an element that are characteristic of that element and that can take part in chemical reactions.

atomic bomb A type of bomb whose explosion is caused by splitting atomic nuclei.

atomic number The number of protons in the nucleus of an atom. The atoms of each element have a different atomic number.

atomic weight The old name for relative atomic mass.

bacteriologist A scientist who studies the nature and behavior of bacteria.

base Any of a group of chemical compounds that, when combined with an acid, form another type of chemical, called a salt, and water.

beta particle An electron, or its positively charged opposite, a positron, emitted from atomic nuclei during some types of radioactive decay.

Big Bang The vast explosion that most scientists believe took place between about 15 and 17 billion years ago which hurled all the matter in the universe outward.

biochemist A scientist who studies the chemical composition of living things.

black hole The collapsed remains of a star with a mass more than three times the sun's. The super-dense mass creates a powerful gravitational field from which no light is able to emerge.

cathode The negatively charged part of electronic devices, such as valves and vacuum tubes, that produces a stream of electrons.

chromosomes 23 pairs of tiny, rod-shaped structures that occur in the nucleus of every human cell. They are made up of thousands of genes, which transmit hereditary characteristics from parents to children.

compound A substance made up of two or more elements whose atoms are joined together chemically.

condensed Changed from a gas to a liquid.

continental drift The gradual movement of the continents over the surface of the Earth.

convection current The constant movement of material below the Earth's surface. Convection currents drive the movement of the Earth's plates.

cosmology The study of the formation, structure, and evolution of the universe.

electrolysis The passing of electricity through a solution or molten solid in order to produce chemical changes.

electromagnetic spectrum The full range of electromagnetic waves, from long-wavelength radio waves to short-wavelength gamma rays.

electron A type of negatively charged particle that is present in all atoms and that orbits the nucleus.

element A substance that contains only one type of atom.

enzyme Any of a group of proteins that occur in cells and speed up the biochemical reactions that take place there.

erosion The gradual wearing away of the Earth's surface by wind, rivers, seas, and ice.

fold A bend in rock that is caused by sideways pressure in the Earth's crust.

fossil The remains of an ancient plant or animal that have been preserved in rock.

galaxy A large group of stars that is held together by gravity and whose members all orbit around a central point. There are four main galaxy types: spiral, barred spiral, elliptical, and irregular.

generator A machine that uses a magnet to produce electricity in wire wrapped around a rotating iron core.

gene One of many thousands of tiny structures that make up chromosomes and that pass characteristics between generations. Different genes carry different characteristics, but all are made of DNA.

genetics The study of the structure and behavior of genes.

gravitational field The area around a body in which the effects of its gravity are experienced.

gravity The force of attraction between objects that have mass.

insulin A hormone produced in an organ of the body called the pancreas, which lies behind the stomach. A shortage of insulin, which controls the amount of sugar in the blood, leads to diabetes.

luminosity The amount of light energy emitted by a star.

magma Hot, molten rock.

magnetic field The area around a magnet in which the effects of its magnetism are experienced.

mass The amount of matter in an object.

matter Any substance that occupies space. Most matter exists as a solid, a liquid, or a gas.

mixture A substance made up of two or more elements whose atoms are simply mixed together, not chemically combined.

molecule The smallest particle of a substance that has its chemical properties. Molecules of elements are made up of two or more chemically joined atoms of the same type. Molecules of compounds are made up of two or more chemically joined atoms of different elements.

motor effect The movement produced in an electric current-carrying wire when it is placed in a magnetic field.

neutron star A dense, collapsed star that has a mass 1.4 to 3 times that of the sun and is made up entirely of neutrons.

neutron A subatomic particle that carries no electric charge, positive or negative. The nuclei of all atoms except hydrogen atoms contain neutrons.

nucleus The positively charged central region of an atom around which negatively charged electrons orbit. Most atomic nuclei are made up of protons and neutrons, but hydrogen nuclei contain only a single proton.

oceanographer A scientist who studies the origin, structure, and development of the oceans and ocean floors.

optics The branch of science that involves the study of vision and light.

Ph.D. The abbreviation for Doctor of Philosophy. This title is granted to people who have gained a doctorate by carrying out original research.

photographic plate A sheet of metal or glass coated with light-sensitive chemicals.

proton A subatomic particle with a positive electric charge. The nuclei of all atoms contain protons, but the number varies. The number of protons in a particular element is known as its atomic number.

radiation The emission of energy from a body in the form of waves or particles. Heat, light, sound, and radioactivity are all forms of radiation.

radio telescope A telescope that is designed to detect radio waves rather than visible light from sources in space.

radioactive decay The emission of particles or rays from the atomic nuclei of elements such as radium as they spontaneously disintegrate.

red supergiant Any of a class of stars that have a higher luminosity than ordinary red giants and are also larger.

relative atomic mass The number of times the mass of an atom of an element is greater or smaller than the mass of one-twelfth of a carbon-12 atom.

rift A huge crack in the Earth's surface caused by two plates moving apart.

sedimentation The settling or depositing of rock, soil, and other particles (sediment) produced during erosion.

solubility The ability of a substance to dissolve in water or another liquid.

sonar An apparatus that detects the position of underwater objects. It does so by sending out sound waves that bounce back when they hit an object, revealing the object's position.

spectroscopy The science of examining light spectra.

spectrum (plural **spectra**) The full range of colors produced when white light is dispersed, for example, by passing it through a prism.

strata (singular **stratum**) Layers, especially layers of rock of different ages and types.

subatomic particle Any of the particles that make up, and are smaller than, atoms. Subatomic particles include electrons, protons, and neutrons.

supernova A star with a mass at least eight times the sun's that collapses once it has burned up all its fuel, then explodes. The explosion causes it to become intensely bright and extremely hot.

INDEX